Artificial Intelligence in Healthcare

LICENSE, DISCLAIMER OF LIABILITY, AND LIMITED WARRANTY

By purchasing or using this book and companion files (the "Work"), you agree that this license grants permission to use the contents contained herein, including the disc, but does not give you the right of ownership to any of the textual content in the book / disc or ownership to any of the information or products contained in it. *This license does not permit uploading of the Work onto the Internet or on a network (of any kind) without the written consent of the Publisher.* Duplication or dissemination of any text, code, simulations, images, etc. contained herein is limited to and subject to licensing terms for the respective products, and permission must be obtained from the Publisher or the owner of the content, etc., in order to reproduce or network any portion of the textual material (in any media) that is contained in the Work.

MERCURY LEARNING AND INFORMATION ("MLI" or "the Publisher") and anyone involved in the creation, writing, or production of the companion disc, accompanying algorithms, code, or computer programs ("the software"), and any accompanying Web site or software of the Work, cannot and do not warrant the performance or results that might be obtained by using the contents of the Work. The author, developers, and the Publisher have used their best efforts to ensure the accuracy and functionality of the textual material and/or programs contained in this package; we, however, make no warranty of any kind, express or implied, regarding the performance of these contents or programs. The Work is sold "as is" without warranty (except for defective materials used in manufacturing the book or due to faulty workmanship).

The author, developers, and the publisher of any accompanying content, and anyone involved in the composition, production, and manufacturing of this work will not be liable for damages of any kind arising out of the use of (or the inability to use) the algorithms, source code, computer programs, or textual material contained in this publication. This includes, but is not limited to, loss of revenue or profit, or other incidental, physical, or consequential damages arising out of the use of this Work.

The sole remedy in the event of a claim of any kind is expressly limited to replacement of the book and/or disc, and only at the discretion of the Publisher. The use of "implied warranty" and certain "exclusions" varies from state to state and might not apply to the purchaser of this product.

Artificial Intelligence in Healthcare

Clinical Decision Support and Modeling

Dominic Etli

MERCURY LEARNING AND INFORMATION
Boston, Massachusetts

Copyright ©2025 by Mercury Learning and Information.
An Imprint of DeGruyter Inc. All rights reserved.

This publication, portions of it, or any accompanying software may not be reproduced in any way, stored in a retrieval system of any type, or transmitted by any means, media, electronic display, or mechanical display, including, but not limited to, photocopy, recording, Internet postings, or scanning, without prior permission in writing from the publisher.

Mercury Learning and Information
121 High Street, 3rd Floor
Boston, MA 02110
info@merclearning.com
www.merclearning.com

D. Etli. *Artificial Intelligence in Healthcare: Clinical Decision Support and Modeling*
ISBN: 978-1-50152-377-9

The publisher recognizes and respects all marks used by companies, manufacturers, and developers as a means to distinguish their products. All brand names and product names mentioned in this book are trademarks or service marks of their respective companies. Any omission or misuse (of any kind) of service marks or trademarks, etc. is not an attempt to infringe on the property of others.

Library of Congress Control Number: 2025937131

242526321   This book is printed on acid-free paper in the United States of America.

Our titles are available for adoption, license, or bulk purchase by institutions, corporations, etc.

All of our titles are available in digital format at various digital vendors.

To my family, whose love and support have always been the foundation of my work.

CONTENTS

Preface		*xv*
Acknowledgments		*xix*
Chapter 1: Introduction to AI and Clinical Informatics		**1**
1.1	A Powerful Synergy in Modern Healthcare	1
	AI and Clinical Informatics	2
	Challenges and Considerations	2
	Real-World Implications	2
	Practical Tips	3
1.2	Defining Artificial Intelligence and Clinical Informatics	3
	Reframing the Role of Data and Knowledge	3
	From Assistant to Innovator: AI's Role in Decision-Making	4
	Machine Learning and Deep Learning	4
	Natural Language Processing (NLP)	5
	Computer Vision	5
	Robotics	5
	Implications for Healthcare Innovation	5
	Systems-Level Impact: Organizational and Ethical Dimensions	6
	Looking Ahead: Ethical and Philosophical Reckonings	7
	Clinical Informatics	7
1.3	A Brief History of AI in Healthcare	8
1.4	The Promise and Potential of AI-Enhanced Care	10
	Improved Diagnostic Accuracy	11

Advancements in Medical Imaging	12
Early Detection and Disease Screening	12
Beyond Imaging: Multimodal Diagnostic Insights	13
Collaborative Approach with Clinicians	13
Regulatory Approvals and Real-World Deployment	13
Challenges and Future Directions	14
Personalized Treatment Plans	14
AI-Driven Treatment Optimization	15
Integrating Multiple Data Streams	16
Case Examples and Practical Applications	16
Ethical and Practical Considerations	17
Future Directions	17
Enhanced Clinical Decision Support	19
Real-Time Alerts and Recommendations	19
Evidence-Based Medicine at the Point of Care	20
Multimodal Data Integration	20
Clinician Acceptance and Workflow Integration	20
Continuous Learning and Feedback Loops	21
Future Directions and Challenges	21
Streamlined Administrative Tasks	23
Automating Clinical Documentation	23
Efficient Scheduling and Appointment Management	24
Revenue Cycle Management and Billing Optimization	24
Predictive Resource Allocation	24
Addressing Quality Assurance and Compliance	25
Challenges and Future Directions	25
Advanced Population Health Management	27
Predictive Modeling for Disease Outbreaks	27
Targeting Preventive Interventions and Chronic Disease Management	28
Addressing Social Determinants of Health	28
Optimizing Resource Allocation and Health System Planning	29

　　　　Ethical and Privacy Considerations　　　　　　　　　　29
　　　　Future Directions in Population Health AI　　　　　　　30

Chapter 2: Machine Learning in Healthcare　　　　　　　**31**
　2.1　The Building Blocks of Intelligent Systems　　　　　　31
　2.2　Machine Learning: Teaching Computers to Learn　　　32
　　　2.2.1　Supervised Learning　　　　　　　　　　　　38
　　　2.2.2　Common Algorithms in Supervised Learning　41
　　　2.2.3　Unsupervised Learning　　　　　　　　　　　44
　　　2.2.4　Reinforcement Learning　　　　　　　　　　47
　　　2.2.5　Clinical Applications of Machine Learning　　48

Chapter 3: Neural Networks and Deep Learning　　　　**55**
　3.1　Neural Networks　　　　　　　　　　　　　　　　　55
　3.2　The Structure of Neural Networks in Healthcare Applications　56
　3.3　Deep Learning Architectures in Healthcare　　　　　57
　3.4　Clinical Implications of Deep Learning　　　　　　　59
　3.5　Programming a Simple Neural Network for Healthcare　61
　3.6　Future Directions　　　　　　　　　　　　　　　　　69

Chapter 4: Natural Language Processing　　　　　　　**71**
　4.1　The Transformer Architecture　　　　　　　　　　　71
　　　4.1.1　Decoder Architecture　　　　　　　　　　　　72
　4.2　Masked Multi-Head Self-Attention　　　　　　　　　74
　4.3　Position-wise Feedforward Layer　　　　　　　　　75
　4.4　Residual Connections and Layer Normalization　　　77
　4.5　Positional Encoding　　　　　　　　　　　　　　　　79
　4.6　Rotary Positional Embedding (RoPE)　　　　　　　82
　4.7　Efficient Attention with Alibi　　　　　　　　　　　84
　4.8　Toward Transformers for Vision　　　　　　　　　　84
　4.9　Multimodal Transformers　　　　　　　　　　　　　85
　　　4.9.1　Text Mining and Information Extraction　　　85
　　　Core Clinical Text Mining Tasks　　　　　　　　　　86
　　　Concept Normalization　　　　　　　　　　　　　　93

		Temporal Information Extraction	98
		Combining Techniques: A Clinical NLP Pipeline	109
	4.9.2	Sentiment Analysis and Opinion Mining	118
4.10	NLP in Clinical Documentation		120
		Important NLP Applications in Clinical Documentation	121
		Advanced Techniques for Clinical Documentation NLP	153
		Challenges and Considerations	154
		Best Practices for Clinical Documentation NLP	156
		The Future of NLP in Clinical Documentation	156

Chapter 5: Data Mining and Predictive Analytics — 159

5.1	Association Rule Mining		160
	5.1.1	Theoretical Foundation	160
	5.1.2	Association Rule Mining Algorithms	160
	5.1.3	Healthcare Applications of Association Rule Mining	161
	5.1.4	Implementation of Association Rule Mining	161
	5.1.5	Interpreting Association Rules in Healthcare	179
	5.1.6	Challenges and Limitations	180
	5.1.7	Advanced Techniques for Healthcare Association Rule Mining	180
	5.1.8	Best Practices for Association Rule Mining in Healthcare	180
5.2	Clustering and Segmentation		181
	5.2.1	Clinical Applications of Clustering	184
	5.2.2	Types of Clustering Algorithms	184
	5.2.3	Implementing Clustering for Patient Segmentation	185
	5.2.4	Clinical Applications of Clustering Results	231
	5.2.5	Challenges and Considerations in Healthcare Clustering	232
	5.2.6	Best Practices for Healthcare Clustering	233
	5.2.7	Future Directions	233
5.3	Predictive Modeling in Patient Care		234

Chapter 6: Core Areas of Clinical Informatics — 237

- 6.1 Clinical Informatics — 237
- 6.2 The Evolution of EHRs — 238
  - 6.2.1 AI-Driven EHR Optimization — 239
  - 6.2.2 Challenges and Future Directions — 241
- 6.3 The Rise of Telemedicine — 245
  - 6.3.1 AI in Remote Patient Monitoring — 246
  - 6.3.2 Overcoming Barriers to Adoption — 248
- 6.4 The Role of CDSSs in Patient Care — 251
  - 6.4.1 AI Techniques in CDSSs — 253
  - 6.4.2 Evaluating CDSS Effectiveness — 277
- 6.5 Data Governance and Regulatory Compliance — 279
  - 6.5.1 AI in Data Protection and Anonymization — 281

Chapter 7: From Bench to Bedside: Translating AI into Practice — 289

- 7.1 From Bench to Bedside: Translating AI into Practice — 289
- 7.2 Imaging Analysis and Interpretation — 292
  - 7.2.1 Pathology and Lab Result Processing — 296
- 7.3 Rare Disease Identification — 302
  - 7.3.1 Personalized Treatment Recommendations — 307
- 7.4 Drug Discovery and Repurposing — 312
- 7.5 Surgical Planning and Assistance — 321
- 7.5 Readmission and Complication Prediction — 329
  - 7.5.1 Chronic Disease Progression Modeling — 337
  - 7.5.2 Social Determinants of Health Analysis — 343
- 7.6 Technical Barriers and Interoperability Issues — 349
- 7.7 Workflow Disruption and User Acceptance — 358
  - 7.7.1 Cost and Resource Constraints — 360
  - 7.7.2 Bias and Fairness in AI Models — 362
- 7.8 Explainability and Transparency — 367
  - 7.8.1 Patient Privacy and Data Ownership — 374

Chapter 8: Case Studies in Clinical Decision Support Systems — 383

- 8.1 Learning from the Pioneers: Real-World Examples — 383
- 8.2 Case Study 1: AI-Powered Diagnostic Decision Support for Rare Diseases — 384
  - 8.2.1 Highlights — 393
- 8.3 Case Study 2: Personalized Treatment Recommendation System for Chronic Diseases — 394
- 8.4 Case Study 3: Real-Time Patient Monitoring and Alert System for Post-Surgical Care — 405
- 8.5 Case Study 4: Clinical Workflow Optimization in Emergency Departments — 418
- 8.6 Lessons Learned and Future Directions — 431

Chapter 9: The Road Ahead: What's Next for AI in Healthcare — 433

- 9.1 Quantum Computing and Healthcare — 436
- 9.2 Edge Computing and Real-Time Analysis — 437
  - 9.2.1 5G Networks and Telemedicine — 455
- 9.3 FDA Guidance on AI/ML-Based Software — 455
  - 9.3.1 GDPR and International Data Protection Laws — 457
- 9.4 Reimbursement Models and Incentive Structures — 459
- 9.5 Federated Learning and Privacy-Preserving AI — 460
- 9.6 Explainable AI and Interpretable Models — 463
- 9.7 Multimodal Learning and Sensor Fusion — 465

Chapter 10: Resources and Continuing Education — 469

- 10.1 The Importance of Continuous Learning — 469
- 10.2 Strategies for Keeping Up-to-Date — 470
- 10.3 Curating Your Information Sources — 471
- 10.4 Recommended Reading and Reference Materials — 472
  - 10.4.1 Important Journals and Publications — 472
  - 10.4.2 Online Articles and Blogs — 473
- 10.5 Conferences and Professional Organizations — 473
  - 10.5.1 Major Healthcare AI Conferences — 473

		10.5.2	Clinical Informatics Societies	474

		10.5.2	Clinical Informatics Societies	474
		10.5.3	Online Forums and Communities	474
	10.6	Training and Certification Programs		475
		10.6.1	University Courses and Degrees	475
		10.6.2	Online Learning Platforms	476
		10.6.3	Professional Certification Bodies	476
	10.7	Putting It All Together: Creating Your Learning Plan		477
		10.7.1	Assessing Your Current Knowledge and Skills	477
		10.7.2	Setting Learning Goals and Objectives	478
		10.7.3	Crafting a Personalized Learning Roadmap	478
	Bibliography			480
Index				**497**

PREFACE

As we begin our exploration of AI in clinical informatics, it's important to understand how this book is structured and how to get the most out of it. This section will guide you through the book's organization and offer tips for maximizing your learning experience.

This book is designed to provide a comprehensive overview of AI in clinical informatics, progressing from foundational concepts to advanced applications and future trends. Here's a brief overview of what you can expect in each chapter:

1. Introduction to AI and Clinical Informatics (current chapter)
2. Foundational Technologies in AI
3. Core Areas of Clinical Informatics
4. Natural Language Processing
5. Predictive Analytics
6. Core Areas of Clinical Informatics
7. Integrating AI into Clinical Workflows
8. Case Studies
9. Future Directions and Emerging Trends
10. Resources and Continuing Education

Each chapter builds upon the previous ones, but they're also designed to be relatively self-contained for readers who want to focus on specific areas.

KEY FEATURES

Throughout the book, you'll encounter several recurring features designed to enhance your learning:

1. *Learning Objectives*: At the start of each section, we'll outline the key concepts and skills you'll gain.
2. *Practical Tips*: Look for these boxed sections that offer actionable advice for applying concepts in real-world settings.
3. *Case Studies*: Real-world examples illustrate how AI is being implemented in various healthcare contexts.
4. *Ethical Considerations*: These sections highlight important ethical issues related to AI in healthcare.
5. *Technical Deep Dives*: For those interested in more technical details, these sections provide in-depth explanations of AI algorithms and methodologies.
6. *References*: Each chapter includes a comprehensive list of citations, allowing you to explore topics further.

HOW TO USE THIS BOOK

Set Clear Objectives: Before starting each chapter, consider what you hope to learn. This will help you focus on the most relevant information for your needs.

Engage Actively: Don't just read passively. Try to relate the content to your own experiences and consider how you might apply the concepts in your work.

Leverage the Practical Tips: These are designed to bridge theory and practice. Consider how you might implement these suggestions in your own healthcare setting.

Explore the Case Studies: These provide valuable insights into real-world applications and challenges. Think about how similar approaches might work in your context.

Use the References: If a topic particularly interests you, don't hesitate to dive deeper using the provided references.

Discuss and Share: Learning is often enhanced through discussion. Consider forming a study group with colleagues to discuss the concepts and their potential applications.

Revisit and Review: As you progress through the book, revisiting earlier chapters can provide new insights as your understanding deepens.

Consider creating a "learning journal" as you read. Note down key concepts, questions that arise, and ideas for applying AI in your own work. This can serve as a valuable resource for future reference and implementation planning.

Remember, the field of AI in healthcare is rapidly evolving. While this book provides a solid foundation, it's important to stay updated with the latest developments. Chapter 7 provides resources for continuing education and staying current in the field.

I hope this book serves as both an informative guide and a practical tool as you navigate the exciting intersection of AI and clinical informatics.

Acknowledgments

Special thanks go to the reviewers and editors who provided constructive feedback and patience during the drafting process, helping to ensure the accuracy and clarity of the material.

On a personal note, I am grateful to my family and friends for their unwavering support and patience throughout this journey. In particular, I thank my wife Kailey for her encouragement and understanding during the long hours of writing and revision.

Finally, I acknowledge the broader community of researchers, clinicians, and innovators in AI and healthcare whose work has laid the foundation for this book. It is my hope that this work contributes to the ongoing dialogue and advancements in this transformative field.

CHAPTER 1

INTRODUCTION TO AI AND CLINICAL INFORMATICS

1.1 A POWERFUL SYNERGY IN MODERN HEALTHCARE

Let's consider an example: a radiologist meticulously examining a complex chest X-ray while, in parallel, an AI algorithm scans the same image for subtle indications of disease. Both detect a small nodule, but the AI also highlights an easily missed anomaly in the lung periphery. This scenario encapsulates the remarkable fusion of human clinical expertise with machine-driven intelligence—a partnership that holds immense promise for improving patient outcomes.

Healthcare is undergoing a profound shift as artificial intelligence (AI) is applied in clinical informatics, transforming how clinicians deliver care and how healthcare systems manage data. This convergence is more than a technological upgrade; it represents a reimagining of patient care and medical decision-making in the era of big data. AI's ability to process and interpret vast datasets enables faster, more accurate insights. A recent McKinsey & Company study estimates that AI could generate up to $100 billion in annual value for the US healthcare system (Abernathy et al., 2020). Clinical informatics has revolutionized the way healthcare information is collected, stored, and accessed, with Electronic Health Records (EHRs) forming the backbone of data-driven healthcare. Today, 96% of hospitals in the United States use certified EHR technology, underscoring just how integral clinical informatics has become (Office of the National Coordinator for Health Information Technology, 2021).

AI and Clinical Informatics

When these two fields intersect, they create a powerful ecosystem capable of handling unprecedented volumes of healthcare data while producing precise, actionable insights. AI-driven clinical decision support systems can now offer real-time recommendations tailored to each patient's unique clinical profile. Meanwhile, EHRs are evolving from static repositories into dynamic platforms that leverage predictive analytics. For instance, an EHR might automatically flag patients at risk for hospital readmissions and suggest proactive interventions, reinforcing clinical decisions with data-backed rationale.

Challenges and Considerations

The integration of AI into healthcare is not without obstacles. Ethical and practical dilemmas are common:

- Privacy and security: Patient data is highly sensitive, and safeguarding it is paramount. More sophisticated AI models often require larger datasets, creating tension between data access needs and patient privacy.

- Algorithmic bias: If AI tools are trained on datasets that underrepresent certain populations, they risk perpetuating or amplifying healthcare disparities.

- Workflow integration: AI solutions must fit naturally into the hectic routines of clinical environments. If poorly integrated, even the most advanced tools can fail to gain traction among busy healthcare professionals.

- Human touch: In healthcare, empathy and rapport are critical. Ensuring that AI augments rather than replaces human interaction is essential to maintaining the trust and compassion central to patient care.

Real-World Implications

Throughout this book, we examine how AI and clinical informatics enhance diagnostics, personalize treatment plans, manage population health, and transform medical education. By examining use cases—such as AI-powered radiology workflows, natural language processing for clinical documentation, and predictive analytics for chronic disease management—we show the opportunities and pitfalls that come with this new paradigm. We also explore the ethical frameworks and regulatory guidelines emerging to govern AI's expansion in healthcare, highlighting best practices for responsible deployment.

Practical Tips

As you explore these topics, think about where AI and clinical informatics could address existing challenges in your own healthcare setting. Are there repetitive administrative tasks that can be automated to free up clinician time? Could complex decision-making scenarios—like triaging urgent care patients—benefit from AI-driven risk scoring? Identifying concrete problems that these technologies can solve will enable you to lead (or champion) their implementation in your organization.

By embracing the synergy between AI and clinical informatics, healthcare professionals can find new areas for patient care, ultimately delivering more accurate diagnoses, personalized treatments, and a more efficient health system.

1.2 DEFINING ARTIFICIAL INTELLIGENCE AND CLINICAL INFORMATICS

To understand the relationship between AI and clinical informatics, it is crucial to first define these terms and their roles in healthcare.

Reframing the Role of Data and Knowledge

In many industries, data is fleeting: captured and deployed briefly for marketing or short-term analytics. In healthcare, however, the temporal arc of data is far more extended, and its contextual importance is significantly greater. AI and clinical informatics go beyond merely analyzing patient data to fundamentally recontextualize knowledge. In this paradigm, each patient's electronic health record becomes a node in a larger "learning network," where clinical notes, diagnostic images, laboratory results, and genetic data converge. Every clinical encounter then transforms into a dual-purpose event, serving immediate patient care needs while also generating rich data that can feed back into the system to refine and enhance future clinical decisions.

This change in managing data results in a move away from episodic, retrospective epidemiological studies toward continuous, real-time analysis. Historically, insights into disease trends or treatment efficacy emerged only after months or years of data collection. Now, machine-learning models trained on current EHR data can rapidly detect emerging disease patterns or novel treatment responses, often in near real-time. As a result, the traditional boundary separating research from clinical practice becomes increasingly blurred. Clinicians can, in effect, participate in ongoing research every time they treat a patient by feeding anonymized, relevant data back into AI algorithms that show updated insights.

From a systemic perspective, the implications are vast. Healthcare organizations gain an agile capacity to predict patient surges, allocate resources effectively, and tailor interventions to individual patient profiles. This shift can also accelerate medical discovery, as new patterns and correlations are identified without the typical delays of grant-funded studies or cumbersome institutional review processes. In essence, combining AI and informatics changes every patient encounter from being a stand-alone episode of care to a data point contributing to a perpetually learning, continuously improving health ecosystem.

From Assistant to Innovator: AI's Role in Decision-Making

Traditionally, AI has filled the role of a "clinical assistant" in healthcare, taking on tasks such as scanning medical images, triaging patients, and generating automated reminders. These functions have been quite valuable for reducing clinician workload and enhancing efficiency. However, as AI technologies mature, they can move beyond supporting established workflows; they may become true "innovators," suggesting novel diagnostic categories, care pathways, and treatment protocols that arise from data patterns no human has previously recognized. This approach highlights the breadth of opportunities for AI to shape the future of healthcare.

Machine Learning and Deep Learning

While traditional machine learning methods excelled at classifying patient data (e.g., determining if a tumor is malignant or benign), deep learning models can now reveal hidden structures within extensive, multidimensional datasets. For instance, advanced algorithms can detect nuanced disease subtypes that have not been formally categorized, or generate precise polygenic risk scores that help pinpoint an individual's likelihood of developing certain conditions. By moving beyond simple "assistant" roles, these algorithms could inspire entirely new diagnostic labels—ones that reflect more granular disease pathways and genetic interactions than current medical taxonomies can capture.

Natural Language Processing (NLP)

NLP has already demonstrated its utility in healthcare by converting the free-text content of patient records into structured data. Yet its capabilities extend well beyond transcription. In the near future, NLP could routinely analyze the emotional undertones of clinical conversations to identify issues like patient distress or clinician burnout. On the diagnostic front, NLP algorithms might scan thousands of patient notes to detect atypical linguistic markers that hint at a rare disease—something that might be overlooked by even the most vigilant human reviewer. By uncovering novel textual signals, NLP transitions from a helpful scribe to a new kind of medical detective.

Computer Vision

In fields such as radiology, dermatology, and ophthalmology, AI-powered computer vision has already proven adept at identifying conditions like diabetic retinopathy often earlier and with equal or superior accuracy compared to human examiners. As these models deepen their capacity to learn, they may begin classifying previously unrecognized disease variants or recommending preventive interventions months—if not years—before a human expert could. This forward-thinking approach elevates AI from a diagnostic aide to a groundbreaking innovator capable of reshaping clinical guidelines around early detection and preventive care.

Robotics

Although robotic systems in medicine (e.g., surgical robots) are often lauded for their mechanical precision, adding AI creates an entirely new paradigm of "collaborative intelligence." Future AI-empowered robots could adapt in real time to a surgeon's technique—like an expert assistant who instinctively knows when to adjust an instrument angle or apply more suction. Beyond surgery, AI-driven robots might learn patient-specific nuances in physical therapy or rehabilitation, crafting personalized programs on the fly. This dynamic interplay not only improves patient outcomes but also redefines what a "team" looks like in modern healthcare.

Implications for Healthcare Innovation

As AI technologies evolve from assistants to active engines of discovery, healthcare systems must prepare for a fundamental shift in how clinical care is conceptualized and delivered. Beyond simply automating or accelerating human tasks, AI models may redefine our entire approach to disease taxonomy, treatment planning, and long-term patient management. Instead of solely enhancing processes like image analysis or appointment scheduling, next-generation AI solutions could propose entirely new categories of disease based on previously undetected data patterns, or recommend treatment pathways informed by genetic, behavioral, and social determinants of health.

To fully realize this potential, healthcare organizations need a robust framework for clinical governance that clearly outlines how AI systems are vetted, validated, and monitored. Ethical oversight becomes paramount, particularly when AI-generated insights could dramatically alter a patient's treatment course. This means instituting mechanisms to detect and correct biases in training data, maintain patient privacy, and ensure equitable access to AI-driven care. Equally important is interdisciplinary collaboration. Clinicians, data scientists, ethicists, and patient representatives must jointly shape guidelines for deploying and scaling these innovations.

By embracing AI as an active partner in innovation, healthcare stakeholders can transcend traditional boundaries. The synergy is no longer confined to expediting existing workflows; it has the power to reshape how diagnoses are defined, how treatment plans are formulated, and ultimately how we conceive the continuum of care. In this emergent paradigm, every decision becomes an opportunity for continuous learning, every encounter a chance to refine our understanding of health and disease. If stewarded responsibly, AI stands to usher in an era of medicine characterized by more personalized, proactive, and precise interventions—profoundly altering not just the present, but also the future trajectory of clinical practice.

Systems-Level Impact: Organizational and Ethical Dimensions

The rapid infusion of AI into clinical informatics is transforming healthcare at multiple levels, causing concerns about workforce displacement, data security, and the potential exacerbation of biases. Yet when used responsibly, AI also fosters a more integrated and proactive health system. Below are several areas in which AI can have a broad, systems-level impact:

1. Equity in care. Properly designed AI algorithms have the potential to identify underserved populations earlier, highlighting gaps in access or flagging disparities that might otherwise remain invisible. By using diverse training data and monitoring for algorithmic bias, healthcare organizations can leverage AI to direct resources more effectively and reduce long-standing inequities in patient outcomes.

2. Workflow evolution. As AI becomes more deeply ingrained in clinical settings, new roles and skill sets are emerging. "Clinician-data scientists" or "health AI interpreters" bridge the gap between raw algorithmic outputs and the nuanced, empathetic care patients expect. These professionals ensure that data-driven insights are translated into actionable, patient-centered decisions, supporting a smoother integration of AI into daily practice.

3. Ethical stewardship. In an era of unprecedented data availability, clear lines of accountability and strong privacy regulations are essential. Traditional notions of patient autonomy and data ownership must be revisited to maintain public trust. Robust ethical frameworks can help mitigate risks related to biased datasets, unintended consequences, or data misuse, reinforcing patient safety and confidentiality.

AI is quickly becoming an inflection point that drives hospitals, payers, and regulators to reimagine long-standing policies, restructure clinical teams, and adopt more transparent approaches to patient engagement and consent. By proactively addressing these organizational and ethical dimensions, healthcare systems can leverage AI not only to enhance efficiency, but also to promote greater equity, innovation, and trust in the delivery of care.

Looking Ahead: Ethical and Philosophical Reckonings

In adopting AI, we must maintain rigorous evaluation of outcome metrics. Does the algorithm *truly* improve patient quality of life, or does it simply reduce wait times? Are we shifting clinician burden from one domain (manual tasks) to another (overseeing AI)? How do we preserve the intrinsic "human touch" of care? It is worth questioning whether we are measuring the "right things" in clinical care. AI's potential for augmenting knowledge is enormous, but it also demands a careful philosophical vantage point: patients are not mere "data points." They are individuals with unique stories, contexts, and clinical complexities that cannot be fully captured by even the most advanced neural networks.

Clinical Informatics

Clinical informatics is the application of informatics and information technology to deliver healthcare services. It involves the use of data and knowledge for scientific inquiry, problem-solving, and decision-making to improve human health and delivery of healthcare services (Kulikowski et al., 2012).

Key components of clinical informatics include the following:

1. Electronic Health Records (EHRs): Digital versions of patients' paper charts. EHRs facilitate the sharing of patient information among healthcare providers and enable data-driven decision support (Cowie et al., 2017).

2. Clinical Decision Support Systems (CDSSs): Tools that provide clinicians with patient-specific assessments or recommendations. CDSSs can alert clinicians to potential drug interactions, suggest evidence-based treatments, and help manage chronic diseases (Sutton et al., 2020).

3. Health Information Exchange (HIE): The electronic movement of health-related information among organizations. HIE allows for the secure sharing of patient data across different healthcare settings, improving care coordination and reducing duplicative tests (Hersh et al., 2015).

4. Telemedicine: The use of technology to deliver care at a distance. Telemedicine can increase access to care, especially in rural or underserved areas, and has proven crucial during the COVID-19 pandemic (Wosik et al., 2020).

1.3 A BRIEF HISTORY OF AI IN HEALTHCARE

The integration of AI in healthcare has been a gradual process, marked by significant milestones and evolving alongside advancements in computing power and data availability. Let's trace this journey through key developments:

1950s-1960s: "The Dawn of AI." The concept of AI emerged in the 1950s, with the term "Artificial Intelligence" coined at the Dartmouth Conference in 1956. Early AI systems in healthcare were rule-based, designed to mimic human decision-making processes. The Ledley-Lusted model, developed in 1959, was one of the first attempts to apply computational reasoning to medical diagnosis (Ledley & Lusted, 1959).

1970s: Expert Systems. The development of medical expert systems occurred in the 1970s. MYCIN, created at Stanford University, was one of the first AI systems designed to identify bacteria causing severe infections and recommend antibiotics. While never used in clinical practice due to legal and ethical concerns, MYCIN laid the groundwork for future clinical decision support systems (Shortliffe, 1976).

1980s-1990s: Machine Learning and Neural Networks. The 1980s and 1990s brought advancements in machine learning algorithms. Neural networks began to show promise in medical image analysis and diagnosis. In 1998, a neural network model demonstrated the ability to predict breast cancer from mammographic findings and patient history, showcasing the potential for AI in cancer detection (Baker et al., 1998).

2000s: Data Mining and Electronic Health Records. The widespread adoption of EHRs in the 2000s provided vast amounts of digital health data. This spurred the development of data mining techniques to extract meaningful patterns from clinical data. In 2001, the Informatics for Integrating Biology & the Bedside (i2b2) project was launched, enabling the use of EHR data for research and clinical decision support (Murphy et al., 2010).

2010s: Deep Learning Revolution. The 2010s marked a significant leap in AI capabilities with the advent of deep learning. In 2012, a deep learning algorithm achieved record-breaking accuracy in the ImageNet Large Scale Visual Recognition Challenge, demonstrating the potential of deep neural networks in image classification (Krizhevsky et al., 2012). This breakthrough paved the way for numerous applications in medical imaging.

2020s: AI in the Clinical Workflow. Currently, there is an increased integration of AI into clinical workflows. AI is being used for everything from triaging radiological scans to predicting patient deterioration in intensive care units. The COVID-19 pandemic has further accelerated the adoption of AI in areas like disease surveillance, vaccine development, and remote patient monitoring (Syeda et al., 2021).

Looking ahead, we can expect AI to play an increasingly central role in healthcare, with a focus on personalized medicine, real-time decision support, and predictive analytics. However, challenges remain, particularly in areas of data privacy, algorithmic bias, and clinical validation.

Practical Tip: When implementing AI solutions, consider the historical context. Many of today's challenges, such as clinical integration and user acceptance, were also a challenge for earlier systems. Learning from past successes and failures can inform more effective implementation strategies.

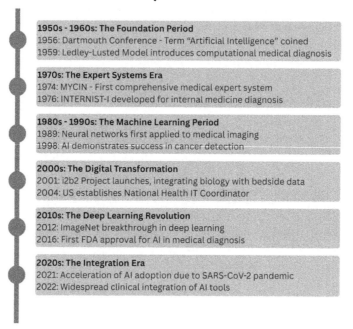

FIGURE 1.1 A brief history of AI in healthcare

1.4 THE PROMISE AND POTENTIAL OF AI-ENHANCED CARE

AI has evolved from a field of theoretical research into an important technology underpinning many real-world applications. In clinical decision-making, AI's capabilities are particularly compelling, given the complexity of medical data and the urgent need for more efficient, personalized, and evidence-based care. By integrating AI into clinical informatics, providers and researchers can use powerful computational techniques—ranging from machine learning and natural language processing to computer vision and robotics—to address issues that have long challenged traditional healthcare systems.

AI's promise lies in its ability to sift through massive, diverse datasets—such as EHRs, medical images, genomic sequences, and patient-reported outcomes—and uncover patterns or trends that would be impossible for humans to detect alone. These patterns, once validated, can inform diagnostic procedures, guide treatment decisions, and optimize resource allocation. Importantly, AI-assisted insights are not meant to replace clinical expertise;

rather, they serve as valuable decision support tools that enhance a clinician's ability to deliver high-quality, patient-centered care.

Efficiency and outcomes are important as healthcare increasingly focuses on value-based models. AI has the potential to streamline administrative workflows, reduce physician burnout, and enable more proactive approaches to population health, such as anticipating disease outbreaks or identifying high-risk patients in real time. Policymakers and clinical leaders emphasize the need for responsible AI implementation, highlighting the importance of robust validation studies, adherence to ethical frameworks, and patient data privacy safeguards. Achieving these standards requires interdisciplinary collaboration among data scientists, clinicians, ethicists, and technology vendors, ensuring that AI-driven tools are transparent, equitable, and safely integrated into existing care pathways.

In the following subsections, we explore how AI is reshaping various domains of healthcare, from improving the accuracy of diagnostic imaging to personalizing treatment plans, enhancing clinical decision support, automating administrative tasks, and facilitating advanced population health management. Each of these applications carries its own set of promises and challenges. Yet taken together, they show the transformative potential AI for reimagining the delivery of patient care, guiding medical research, and ultimately improving health outcomes on a global scale.

Improved Diagnostic Accuracy

AI algorithms, particularly those utilizing deep learning techniques, have demonstrated their capabilities in medical imaging analysis and other diagnostic tasks. By extracting complex features that may be imperceptible to the human eye, these algorithms can improve the speed, sensitivity, and specificity of disease detection, thereby enhancing the overall diagnostic accuracy of healthcare systems.

Advancements in Medical Imaging

One of the most widely recognized applications of AI in healthcare is in the field of medical imaging. In a 2019 study published in *The Lancet Digital Health*, researchers reported that deep learning algorithms achieved a specificity of 92.0% and a sensitivity of 86.9% in detecting breast cancer from mammograms, surpassing the performance of human radiologists (McKinney et al., 2020). Similar breakthroughs have been documented in the following areas:

- Dermatology: AI-driven image classifiers can identify malignant skin lesions with accuracies comparable to experienced dermatologists (Esteva et al., 2017).
- Pulmonology: Deep learning models have shown promise in detecting early stage lung cancer nodules on CT scans (Ardila et al., 2019).
- Ophthalmology: Algorithms developed to analyze retinal images can diagnose diabetic retinopathy with high sensitivity and specificity (Gulshan et al., 2016).

What makes AI-driven imaging solutions particularly compelling is their ability to learn continuously. As these models process larger and more diverse datasets, they refine their detection capabilities, improving performance over time. This iterative learning process helps ensure that AI tools stay current with evolving disease patterns and imaging protocols.

Early Detection and Disease Screening

The potential for AI to *facilitate earlier disease detection* cannot be overstated. Many conditions, including cancers and chronic diseases, are most treatable at an early stage. By flagging subtle imaging abnormalities or laboratory values that would otherwise go unnoticed, AI can prompt clinicians to initiate timely interventions. This early detection advantage is especially valuable in population-level screening programs, where large volumes of images must be interpreted quickly and accurately.

AI can help prioritize urgent cases in radiology workflows, reducing wait times for patients who require immediate attention. In busy clinical settings, automated triage systems can sift through incoming scans—such as chest X-rays—to highlight possible anomalies for rapid evaluation. This approach ensures that patients with critical pathologies do not face unnecessary delays in diagnosis.

Beyond Imaging: Multimodal Diagnostic Insights

While medical imaging has garnered the most attention, AI-driven diagnostic improvements extend well beyond radiology. *Clinical decision support systems* can integrate data from multiple sources—lab results, genomic information, patient-reported outcomes, and EHR metadata—to offer a more holistic view of a patient's condition. By synthesizing and analyzing multimodal data, AI tools can

- identify *patterns or correlations* across different diagnostic tests
- predict *risk scores* for adverse events or disease progression
- suggest *personalized diagnostic pathways*, tailoring the next set of investigations based on individual risk factors

This integrated approach helps clinicians make *data-informed decisions* that can shorten diagnostic timelines and reduce the risk of costly or invasive tests.

Collaborative Approach with Clinicians

Importantly, AI does not replace clinicians but rather augments their expertise. By taking on time-intensive tasks such as scanning large volumes of images or cross-referencing medical records, AI frees radiologists and other specialists to focus on complex decision-making and patient communication. In practice, many institutions are moving toward *hybrid models* wherein AI tools and expert clinicians collaborate—often referred to as "AI-assisted diagnostics." This synergy combines the speed and pattern-recognition strengths of algorithms with the nuanced judgment and clinical context that only human professionals can provide.

Additionally, studies have shown that *human-AI collaboration* often yields better outcomes than AI or human interpretation alone. For instance, an AI tool might detect suspicious regions in an MRI scan, while a radiologist applies knowledge of the patient's history and physical exam findings to interpret the significance of those findings, ultimately arriving at a more accurate diagnosis.

Regulatory Approvals and Real-World Deployment

As AI algorithms transition from research labs to clinical settings, regulatory bodies such as the US Food and Drug Administration (FDA) have begun to establish pathways for clearance and approval of AI-based diagnostic tools. These frameworks address issues such as algorithmic transparency, data privacy, and patient safety. Although the regulatory landscape is still evolving, the increasing number of AI tools obtaining clearance signals a growing confidence in their real-world utility.

Implementation in actual healthcare workflows also involves robust validation across diverse patient populations to confirm that AI models perform reliably outside the controlled environment of a clinical trial or retrospective data study. Careful monitoring of model performance after deployment is critical to detect any drift in accuracy, especially as new types of data or imaging devices enter the pipeline.

Challenges and Future Directions

Despite the rapid improvements in AI-driven diagnostic accuracy, several challenges remain:

- Data quality and bias: Models trained on limited or unrepresentative datasets may underperform in different demographic or clinical settings. Ensuring high-quality data is essential.
- Explainability: Many deep learning approaches are considered "black boxes," making it difficult for clinicians to understand how a particular conclusion was reached. Emerging methods in explainable AI aim to address this gap.
- Ethical and legal considerations: As AI tools gain influence in clinical decision-making, questions around liability, ethics, and accountability become more pressing.
- Integration and workflow: Achieving seamless integration with existing health IT systems and ensuring clinical adoption remain key hurdles for widespread AI acceptance.

Looking ahead, AI-driven diagnostic tools are expected to become even more sophisticated, incorporating NLP to analyze clinical notes alongside imaging, lab, and genomic data (Sendak et al., 2019). Additionally, advances in federated learning may allow models to train collaboratively on multiple healthcare institutions' data without compromising patient privacy. As these innovations unfold, clinicians, data scientists, and health system leaders must work together to ensure that improved diagnostic accuracy translates into tangible health benefits for diverse patient populations.

Personalized Treatment Plans

Personalized medicine, often used synonymously with *precision medicine*, attempts to optimize healthcare by tailoring treatment strategies to the unique characteristics of each patient (Schork, 2019). These characteristics may include genetic makeup, molecular profiles, lifestyle factors, and environmental exposures. AI enables the integration and analysis of diverse data streams to form a comprehensive picture of patient health, resulting in more precise and effective treatment decisions.

Leveraging Genetic and Genomic Information

An important part of personalized medicine is the use of genetic and genomic data:

- Pharmacogenomics: AI can sort through large genomic databases to identify genetic variants that influence drug metabolism. This allows clinicians to predict which medications or dosages will be most effective and least likely to cause adverse reactions.

- Cancer genomics: Certain tumors exhibit distinct genetic signatures that inform targeted therapies. By analyzing genomic markers, AI systems can suggest the most promising treatments—for instance, immunotherapies or specific kinase inhibitors tailored to the tumor's mutation profile.

Such analyses often rely on machine learning models that can discern subtle genomic patterns, correlating them with treatment responses and patient outcomes. As these models absorb increasingly robust datasets, they become better at matching individual patients to therapies that maximize efficacy while minimizing side effects.

AI-Driven Treatment Optimization

While genetic data remains essential, personalized treatment also involves dynamic factors like a patient's comorbidities, medication history, and real-time physiological data. AI algorithms excel at combining these heterogeneous data points to deliver more nuanced recommendations than could be achieved through traditional clinical guidelines alone.

Machine learning and deep learning methods can do the following:

- Identify optimal dosages. In a 2021 study published in *Nature Medicine*, researchers used machine learning to predict personalized radiation therapy doses for cancer patients, optimizing treatment efficacy and minimizing adverse effects (Kang et al., 2021). By accounting for tumor volume, patient anatomy, and past treatment responses, the AI approach offered an individualized radiation plan superior to one-size-fits-all protocols.

- Predict treatment outcomes. AI can forecast patient responses to various therapies, such as chemotherapy regimens or combination drug treatments, based on a patient's clinical profile. These predictions help guide clinicians toward therapies with the highest likelihood of success (Ribeiro, 2016).

Integrating Multiple Data Streams

Personalized treatment plans are most useful when they incorporate a broad spectrum of data:

- Electronic Health Records (EHRs): Demographics, clinical notes, lab results, and medication histories can be mined to reveal patterns that correlate with treatment success or failure.
- Lifestyle and wearable data: Patient-generated data from fitness trackers or wearable sensors can alert clinicians to daily fluctuations in vital signs or activity levels. AI can integrate these signals to adjust treatments in near real time (Etli et al., 2024).
- Socioeconomic and environmental factors: Social determinants of health, such as living conditions, access to nutritious food, and community resources, significantly impact treatment adherence and outcomes. AI-driven models can incorporate these contextual elements to ensure that proposed treatment plans are both effective and feasible.

By unifying clinical, genomic, behavioral, and environmental data, personalized treatment models can provide holistic recommendations that align with a patient's needs and circumstances (Ramesh et al., 2013).

Case Examples and Practical Applications

- Oncology: Beyond radiation therapy dosing, AI-driven platforms help oncologists choose the most promising chemotherapy regimens based on tumor genomics. For instance, machine learning models can predict which patients are likely to develop resistance to specific drugs, guiding timely therapy switches.
- Cardiology: Personalized risk stratification models identify which patients might benefit from early interventions such as statins, beta-blockers, or lifestyle modifications to prevent heart failure.
- Mental Health: AI-powered decision support systems can match individuals to counseling methods or medications based on historical treatment data, reducing the trial-and-error typically involved in psychiatric care.

Ethical and Practical Considerations

- Data privacy and security: Personalized treatment plans often involve sensitive genetic information and other private health details. Ensuring secure data storage and adherence to privacy regulations (e.g., HIPAA) is important.

- Equitable access: Advanced genetic testing and AI-driven analytics can be expensive, raising concerns about unequal access. Policymakers and healthcare organizations must consider reimbursement models and patient support programs to avoid exacerbating health disparities.

- Model explainability: As AI tools become more integral to clinical decision-making, clinicians and patients need explanations for why a particular treatment is recommended. Explainable AI techniques can help build trust and facilitate shared decision-making.

- Regulatory oversight: Personalized treatments often require approval from regulatory bodies, especially when algorithms suggest off-label drug usage or novel therapeutic pathways. Ongoing collaboration between AI developers, clinicians, and regulators is needed to define safe, evidence-based processes for integrating AI into clinical practice.

Future Directions

Looking ahead, the field of personalized treatment is poised for ongoing innovation:

- Continuous learning health systems: Healthcare institutions are developing integrated data ecosystems where each new patient case further refines AI models, fostering adaptive improvement in treatment recommendations.

- Multi-omics integration: Beyond genomics, future AI-driven systems will combine transcriptomics, proteomics, and microbiomics data to gain a deeper understanding of disease mechanisms and refine treatment regimens.

- Real-time personalization: Wearable biosensors, implantables, and telemedicine platforms make it possible for real-time, AI-supported adjustments to therapies—such as insulin pumps that automatically adjust dosages based on continuous glucose monitoring.

Ultimately, personalized treatment plans that utilize AI are part of a paradigm shift in medicine. By moving from a reactive to a proactive model of care, clinicians can address the unique aspects of each patient's biology and life circumstances, thereby improving outcomes, enhancing the patient experience, and reducing unnecessary interventions. As AI tools evolve, their capacity to synthesize vast amounts of information will continue to expand, offering ever more refined and patient-centered treatment strategies.

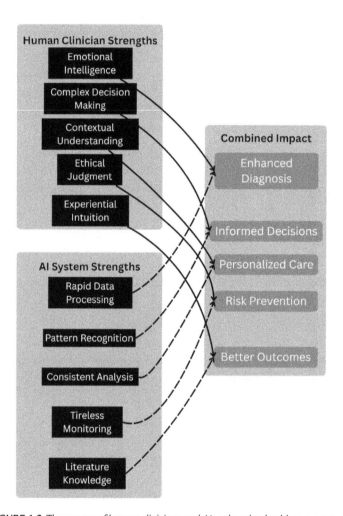

FIGURE 1.2 The synergy of human clinicians and AI: enhancing healthcare outcomes

Figure 1.2 illustrates the synergy between human clinicians and AI systems, showcasing how their combined strengths lead to improved healthcare outcomes. "Human Clinician Strengths" include emotional intelligence, complex decision-making, contextual understanding, ethical judgment, and intuition from experience. "AI System Strengths" encompass rapid data processing, pattern recognition, consistent analysis, tireless monitoring, and literature knowledge. The center section shows how these strengths integrate, leading to "Combined Impact:" an enhanced diagnosis, informed decisions, personalized care, risk prevention, and ultimately, better outcomes.

Enhanced Clinical Decision Support

Clinical Decision Support Systems (CDSSs) are integral to modern healthcare, providing clinicians with evidence-based insights that can reduce errors, streamline workflows, and bolster patient outcomes. AI has accelerated the changes in CDSSs by bringing new levels of automation, predictive accuracy, and real-time analysis to bedside decision-making. In a 2020 systematic review published in *BMJ Health & Care Informatics*, researchers found that AI-based CDSSs improved patient outcomes in 64% of the studies reviewed (Muehlematter et al., 2021). As these systems continue to mature, they offer significant potential to further elevate the quality and efficiency of care.

Real-Time Alerts and Recommendations

One of the strengths of an AI-integrated CDSS is the ability to deliver actionable insights in real time. Rather than having clinicians search the medical literature or patient data manually, AI systems rapidly analyze EHRs, lab results, and imaging data to

- flag potential drug interactions: By cross-referencing a patient's medication list with known interactions, the system can immediately alert prescribers to potentially dangerous combinations.

- recommend diagnostic tests: Based on presenting symptoms and patient history, AI can suggest additional labs or imaging studies that may confirm or rule out specific conditions.

- predict patient deterioration: Predictive algorithms can detect early warning signs—e.g., subtle changes in vital signs or lab trends—enabling clinicians to intervene before a severe event occurs.

In high-acuity settings such as intensive care units (ICUs) and emergency departments, these real-time alerts often make the difference between timely, life-saving interventions and delayed responses.

Evidence-Based Medicine at the Point of Care

AI-driven CDSS platforms frequently incorporate machine learning models trained on large-scale clinical datasets, guidelines, and research studies. This integration connects cutting-edge research and day-to-day clinical practice by

- summarizing the latest evidence: Systems can present concise, patient-specific summaries of relevant clinical trials or treatment guidelines, saving clinicians hours of literature review.
- tailoring recommendations to local protocols: Using global research, a CDSS can be configured to align with an institution's formulary, diagnostic pathways, and standard operating procedures.

Such "evidence-based recommendations" improve the consistency of care and patient safety by reducing practice variation and ensuring that clinical decisions use the most up-to-date information.

Multimodal Data Integration

Advanced CDSS solutions utilize multimodal data (including imaging, lab tests, continuous vital sign monitoring, genomic information, and social determinants of health) to offer more holistic guidance. By correlating these diverse data points,

- complex diagnoses can be identified earlier. For instance, an AI system might integrate subtle EKG abnormalities with lab markers to raise suspicion for sepsis or acute cardiac events.
- risk stratification becomes more precise. Predictive models can categorize patients into high-, moderate-, or low-risk groups for targeted interventions, such as extra nursing support or specialist consults.

This holistic approach can be especially valuable for complex, multi-morbid patients, those with multiple chronic conditions who might otherwise suffer from the use of conventional decision support tools focused on one disease at a time.

Clinician Acceptance and Workflow Integration

Despite their potential, AI-based CDSSs must gain the trust and acceptance of clinicians to drive meaningful improvements in patient outcomes. Common considerations include the following:

- User-friendly interfaces: Systems should present alerts and recommendations concisely, without overwhelming clinicians with excessive notifications or convoluted dashboards.

- Explainability: Many AI algorithms, particularly deep learning models, operate as "black boxes." Efforts to develop explainable AI solutions help clinicians understand the rationale behind a system's recommendations, fostering confidence in the tool.
- Workflow alignment: To prevent alert fatigue and ensure adoption, AI-driven recommendations must be integrated seamlessly into existing EHRs and clinical pathways. Overly disruptive or time-consuming workflows often lead to underutilization, regardless of the system's accuracy.

Successful change management and training initiatives are vital, ensuring that care teams know how to interpret and act on CDSS outputs.

Continuous Learning and Feedback Loops

AI-based CDSSs do not remain static after deployment. Rather, they improve with continuous learning:

- Monitoring performance: As new patient data are generated, the system's predictions can be compared against actual outcomes. This feedback loop provides opportunities to refine and recalibrate the underlying algorithms.
- Incorporating new evidence: Clinical guidelines evolve with ongoing research, and CDSS tools must be updated accordingly to retain their relevance.
- Adapting to shifting patient populations: Demographic changes, emerging diseases, and new treatment modalities require CDSS models to adjust their parameters and maintain accuracy over time.

These iterative improvements enable AI-based systems to remain agile in dynamic healthcare environments, constantly improving their utility and reliability.

Future Directions and Challenges

The role of AI in CDSS will likely expand as algorithms become more sophisticated and data sources begin to include more types of information. However, several challenges remain:

- Interoperability: Healthcare data are often siloed in disparate systems. Standardization and improved data sharing are crucial for building truly robust, cross-institutional CDSSs.

- Data quality and bias: Biased or incomplete training data can lead to erroneous or inequitable recommendations. Ongoing vigilance is necessary to identify and mitigate biases in algorithmic outputs.

- Regulatory oversight: Government agencies continue to develop frameworks for approving and monitoring AI-based medical devices, including CDSS. Compliance with evolving regulations is essential for safe, widely accepted deployment.

- Ethical considerations: From patient privacy concerns to the potential for overreliance on AI, ethical frameworks must guide the responsible use of advanced CDSS in clinical practice.

Despite these challenges, AI-enhanced clinical decision support represents a pivotal leap in healthcare, enabling clinicians to navigate complex clinical scenarios with greater confidence and precision. Whether by identifying life-threatening conditions earlier or customizing treatments to individual patient profiles, AI-based CDSSs are indispensable in modern medical practice, helping practitioners make more informed, data-driven decisions for the benefit of their patients.

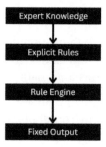

FIGURE 1.3 Traditional rule-based AI: A fixed decision-making approach

This diagram outlines the structure of traditional rule-based AI, which follows a deterministic decision-making process. It starts with expert knowledge, where human expertise defines the foundation. Next, explicit rules are established, forming a structured guideline for decision-making. The rule engine then applies these predefined rules to incoming data. Finally, this system produces a fixed output, meaning the AI follows strict, predefined logic without adaptation or learning from new data. This contrasts with modern machine learning approaches, which rely on dynamic pattern recognition and adaptability.

Streamlined Administrative Tasks

Administrative demands in healthcare often result in cumbersome workflows, high costs, and clinician burnout. From insurance reimbursements to appointment scheduling and clinical documentation, these non-clinical tasks consume substantial time and resources. AI has emerged as a powerful tool to automate or optimize many of these processes, freeing healthcare professionals to dedicate more attention to direct patient care.

Automating Clinical Documentation

One of the most promising applications of AI lies in NLP, which can parse unstructured text (from clinical notes to patient narratives) into structured data. By automating important parts of clinical documentation, NLP helps

- extract relevant information such as symptoms, diagnoses, and medications from physician dictations or free-text patient encounters.
- reduce manual charting time by auto-filling EHR fields with the extracted data.
- decrease transcription errors and enhance the overall quality of medical records.

For instance, a 2019 study in *JMIR Medical Informatics* demonstrated the high accuracy of NLP systems in extracting symptoms, medications, and diagnoses from clinical notes, saving both time and reducing errors (Koleck et al., 2019). Implementing AI-enabled documentation systems can thus alleviate a major source of stress for healthcare providers, who often spend substantial hours each day updating medical charts.

Efficient Scheduling and Appointment Management

Missed appointments, scheduling inefficiencies, and long wait times contribute to patient dissatisfaction and lost revenue. AI scheduling tools help with these issues by

- predicting appointment durations and no-show probabilities using patient demographics, historical attendance, and travel logistics
- recommending optimal appointment slots to minimize patient wait times and balance clinician workload
- automating reminders and follow-ups via text or email, improving patient adherence to scheduled visits

By reducing scheduling complexity and automating labor-intensive tasks, these solutions enhance operational efficiency. Some systems even use reinforcement learning to continually refine scheduling patterns, adjusting in real-time to last-minute cancellations or surges in patient demand.

Revenue Cycle Management and Billing Optimization

Billing and claims processing are pain points in healthcare administration. AI-enabled billing systems can

- identify coding discrepancies by comparing clinical documentation with standard billing codes, helping prevent both overbilling and underbilling
- identify potential errors or incomplete claims before submission, reducing rework and denials from insurance providers
- optimize revenue capture by uncovering overlooked billable items or services

Through automated coding and claims validation, hospitals and clinics can significantly cut down on administrative costs, improve cash flow, and reduce delays in reimbursement.

Predictive Resource Allocation

Beyond documentation and billing, predictive analytics, a subset of AI, enables health systems to optimally allocate resources by forecasting patient volumes and service demands:

- Staffing: By analyzing historical admissions and seasonal trends, AI can help nurse managers and department heads anticipate peak workloads and schedule staff accordingly.
- Bed management: Real-time prediction of patient discharge times and incoming admissions helps ensure that beds are available when needed, preventing bottlenecks in emergency departments and reducing patient wait times.
- Inventory and supply chain: Intelligent systems can track inventory usage patterns for everything from personal protective equipment (PPE) to specialty drugs, automatically prompting reorders and minimizing stockouts.

These data-driven insights ensure that healthcare facilities deploy resources efficiently, reducing costs and enhancing the flow of patients.

Addressing Quality Assurance and Compliance

Maintaining compliance with regulatory standards, such as HIPAA in the US, and ensuring high-quality patient care are critical administrative considerations. AI tools can support continuous quality assurance by

- monitoring for errors or omissions in clinical and billing documentation
- cross-referencing patient outcomes with established care standards to detect deviations
- generating audit reports that help internal review boards and external auditors quickly assess compliance

By proactively identifying potential issues, organizations can rectify them before they escalate, safeguarding both patient safety and institutional reputation.

Challenges and Future Directions

While AI administrative solutions hold tremendous promise, several factors deserve careful attention:

- Integration with legacy systems: Many healthcare institutions rely on EHRs or billing platforms that lack interoperability with AI tools, complicating their implementation.
- Data security and privacy: Automating administrative tasks often involves handling sensitive patient information, which must be securely stored and transmitted.
- Workforce adaptation: Shifting from manual processes to AI-enhanced workflows can meet resistance if staff are not adequately trained or remain skeptical about new technologies.
- Regulatory frameworks: As with other AI applications in healthcare, administrative tools must adhere to evolving standards and guidelines to ensure accurate, fair, and ethical processes.

Streamlined administrative tasks enhanced by AI can further transform healthcare operations. Future innovations may incorporate machine learning algorithms that automatically adapt to changing organizational needs in real time, robotic process automation (RPA) to handle repetitive tasks at scale, and intelligent chatbots to assist with patient inquiries and telemedicine appointments. By offloading administrative burdens, these advances can allow clinicians to focus more on what truly matters: delivering high-quality, compassionate care to their patients.

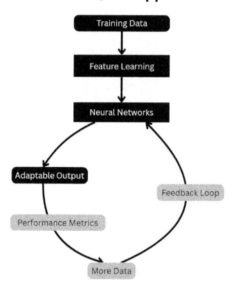

FIGURE 1.4 Modern machine learning and deep learning approaches: A dynamic learning system

Figure 1.4 illustrates the iterative and adaptive nature of modern machine learning (ML) and deep learning (DL) approaches. The process begins with training data, which feeds into feature learning, allowing the model to identify relevant patterns. Neural networks then process this learned information to generate an adaptable output. Unlike traditional rule-based AI, this system continuously improves through a feedback loop, where performance metrics assess outcomes, leading to the collection of more data for further refinement. This cyclical process enhances learning, enabling the model to evolve and improve over time.

Advanced Population Health Management

Population health management aims to improve health outcomes at the community or regional level by focusing on preventive care, effective resource allocation, and equitable access to medical services. AI helps stakeholders, ranging from public health agencies to integrated healthcare systems, analyze vast datasets and gain actionable insights. During the COVID-19 pandemic, for example, AI played a pivotal role in forecasting virus spread, optimizing hospital resources, and expediting drug and vaccine development (Lalmuanawma et al., 2020). Beyond acute crises, AI-driven population health management can help identify high-risk individuals, target preventive interventions, and reduce healthcare disparities in the long term.

Predictive Modeling for Disease Outbreaks

One of the most notable applications of AI in population health is epidemiological modeling:

- Early detection of outbreaks: By continuously parsing data from multiple sources, such as EHRs, symptom-tracking apps, and social media, AI algorithms can detect unusual patterns that signal a brewing outbreak. This allows public health authorities to initiate interventions earlier, potentially containing an epidemic before it escalates.

- Forecasting spread and severity: Machine learning models, including time-series analysis and spatial clustering, can project the trajectory of infectious diseases. These forecasts help policymakers and hospital systems anticipate patient surges, allocate critical resources (e.g., ICU beds and ventilators), and implement targeted containment measures.

During the COVID-19 pandemic, AI-enhanced predictive analytics proved crucial for anticipating local surges, informing social distancing guidelines, and guiding strategic decisions about testing and vaccine distribution.

Targeting Preventive Interventions and Chronic Disease Management

AI can significantly enhance preventive strategies by identifying populations at heightened risk for specific conditions:

- Chronic illnesses: Algorithms can analyze EHR data, medication histories, and demographic factors to identify patients who are prone to uncontrolled hypertension, diabetes complications, or heart failure exacerbations. Care teams can then prioritize these high-risk patients for early interventions, such as telehealth check-ins or home monitoring devices, to prevent hospitalizations.

- Lifestyle and behavioral factors: By integrating wearable sensor data, dietary logs, and physical activity records, AI can identify patients who might benefit from lifestyle modifications. The result is a more proactive, data-driven approach to chronic disease management that empowers patients to take greater control of their health.

These early, targeted measures can shift healthcare from a reactive to a proactive model, reducing both patient morbidity and overall healthcare expenditures.

Addressing Social Determinants of Health

Social determinants of health (SDOH), including economic stability, education, and access to healthcare, are often important predictors of patient well-being and health outcomes. AI tools can help incorporate SDOH data into population health strategies by

- identifying underserved areas: Machine learning models that integrate neighborhood-level socioeconomic metrics (e.g., income, education, housing conditions) can show geographic hotspots where patients lack access to primary care or healthy food options.
- personalizing interventions: AI can recommend community resources—like subsidized meal programs or transportation services—to individual patients who face barriers to healthcare. Clinicians and care coordinators can then tailor follow-up plans based on these social and environmental contexts.

By considering all social and environmental factors, AI-enhanced population health management can reduce health disparities and provide more equitable care.

Optimizing Resource Allocation and Health System Planning

AI also offers robust solutions for managing healthcare resources more efficiently:

- Hospital capacity and staffing: Predictive models that analyze patient flow, seasonal trends, and regional disease prevalence help administrators determine optimal staffing levels and bed capacity. This is especially critical during flu seasons, weather-related crises, or other surges in demand.
- Vaccine and medication distribution: In scenarios like pandemic response or routine immunization drives, AI algorithms can factor in population density, demographic risk factors, and transportation logistics to optimize distribution pathways, ensuring vaccines or essential medications reach the most vulnerable populations first.
- Cost-effectiveness analysis: AI-enhanced analytics can identify the costs and benefits of different preventive programs or therapeutic interventions.

For instance, a health system might use machine learning to compare outcomes for community-based diabetes prevention programs versus more intensive, clinic-based interventions.

Such data-informed planning enables healthcare organizations to allocate limited resources where they can have the greatest impact on community well-being.

Ethical and Privacy Considerations

While AI-driven population health strategies present transformative opportunities, they also raise several ethical and privacy concerns:

- Data security: Large datasets containing personal health information must be encrypted and stored securely to prevent unauthorized access.

- Bias and equity: Models trained on biased or incomplete data may inadvertently reinforce existing inequities, particularly for minority or underrepresented groups. Continuous monitoring and validation are critical to ensure fair and accurate predictions.

- Informed consent: As AI systems aggregate and analyze data from diverse sources (e.g., social media, consumer wearables), questions arise about patient consent and data ownership.

- Transparency: Policymakers, clinicians, and the public need clarity about how AI algorithms operate, especially when they inform decisions about healthcare resource allocation.

Balancing the benefits of large-scale analytics against the need for privacy and fairness is paramount to maintaining public trust and ensuring ethical usage of AI.

Future Directions in Population Health AI

As AI technology continues to evolve, several emerging trends may shape the future of population health management:

- Federated learning: This approach enables multiple organizations or institutions to collaborate on AI model training without sharing patient-level data, preserving privacy while enhancing the robustness of algorithms.

- Real-time dashboards and alerts: Integrated platforms can continuously monitor community health metrics, such as infection rates or hospitalization numbers, offering live dashboards for regional health authorities.

- Cross-sector collaboration: Population health interventions increasingly involve partnerships between healthcare, social services, and government agencies. AI can facilitate data sharing and coordination among these stakeholders, amplifying the overall impact of public health strategies.

- "Precision" public health: Building on the concept of precision medicine at the individual level, precision public health aims to deliver more targeted, population-wide interventions by tailoring them to specific genetic, cultural, or environmental contexts.

Whether it involves predicting the path of a global pandemic or pinpointing local hotspots of chronic disease, AI-enhanced population health management can help create a more proactive and equitable healthcare ecosystem. By identifying critical patterns in large-scale data, streamlining the deployment of interventions, and addressing social determinants of health, AI empowers healthcare leaders to make informed, timely decisions that improve outcomes for communities at large.

Practical Tip: When considering AI implementation in your healthcare setting, start by identifying specific pain points or inefficiencies in your current workflows. AI solutions are most effective when they address concrete, well-defined problems rather than being implemented for their own sake.

CHAPTER 2

MACHINE LEARNING IN HEALTHCARE

2.1 THE BUILDING BLOCKS OF INTELLIGENT SYSTEMS

Artificial intelligence (AI) is revolutionizing the healthcare industry, offering new ways to improve patient care, streamline processes, and support clinical decision-making. However, to fully obtain the potential of AI in healthcare, it is crucial to understand the foundational technologies that make intelligent systems possible.

At its core, AI is about creating machines that can perform tasks that typically require human intelligence, such as learning, problem-solving, and pattern recognition (Russell & Norvig, 2021). This is achieved through a combination of algorithms, data structures, and computing power that enable machines to process and analyze vast amounts of information.

An important building block of AI is *machine learning*, which involves training algorithms to learn from data without being explicitly programmed (Goodfellow et al., 2016). By exposing machine learning models to large datasets, they can automatically identify patterns and relationships that may be too complex or subtle for humans to discern. This enables them to make predictions, classify information, and generate insights that can inform clinical decision-making.

Another important component of AI is neural networks, which are inspired by the structure and function of the human brain (LeCun et al., 2015). Neural networks consist of interconnected nodes that process and transmit information, allowing them to learn and adapt based on the data they receive. Deep learning, a subfield of machine learning, utilizes neural networks with many layers to learn hierarchical representations of data, enabling them to engage in complex tasks such as image recognition and natural language processing.

Natural language processing (NLP) is another essential building block of AI, particularly in healthcare where much of the data is unstructured and text-based (Demner-Fushman et al., 2020). NLP techniques enable machines to understand, interpret, and generate human language, allowing them to extract meaningful information from clinical notes, research papers, and patient narratives. This can help clinicians quickly find relevant information, identify potential risks or complications, and communicate more effectively with patients.

Finally, data mining and predictive analytics are important components of AI that involve discovering patterns and insights from large datasets (Han et al., 2022). By applying statistical and machine learning techniques to healthcare data, these tools can uncover hidden relationships between variables, identify risk factors for diseases, and predict patient outcomes. This can help healthcare organizations allocate resources more effectively, intervene early to prevent complications, and personalize treatment plans based on individual patient characteristics.

While each of these building blocks is powerful on its own, the real potential of AI lies in their integration and synergy. By combining machine learning, neural networks, natural language processing, and data mining, intelligent systems can address complex healthcare challenges and provide valuable insights to support clinical decision-making.

However, it is important to recognize that AI is not a panacea and there are significant challenges and considerations in its application to healthcare. These include issues of data quality and bias, the interpretability and transparency of AI models, and the need for human oversight and judgment in clinical contexts (Topol, 2019). As we explore the clinical applications of AI in the following sections, we keep these challenges in mind and discuss strategies for responsible and effective implementation.

2.2 MACHINE LEARNING: TEACHING COMPUTERS TO LEARN

Machine learning is a fundamental aspect of AI that enables computers to learn from data without being explicitly programmed to do so (Mitchell, 1997). By exposing machine learning algorithms to large datasets, they can automatically identify patterns, extract insights, and make predictions that can inform clinical decision-making. Machine learning has become increasingly important in healthcare as the volume and complexity of medical data continue to grow, making it difficult for clinicians to manually process and analyze all available information (Beam & Kohane, 2018).

Here, let us review some of the foundational aspects of machine learning. There's a concept known as *Bag of Words*. Imagine you have a collection of documents and need to determine the primary topic of each one. This task, when the topics are predefined, is referred to as *classification*. If there are only two possible topics, it falls under the category of binary classification, as discussed in Section 1.7. However, when there are more than two topics to classify, the problem becomes one of multiclass classification.

In multiclass classification, the dataset is represented as a series of paired observations:

$$\{(x_i, y_i)\}_{i=1}^{N},$$

where x_i represents the input data (e.g., a text document), y_i is the corresponding label (e.g., the topic), N is the total number of examples, and C denotes the total number of classes. The label y_i is typically an integer indicating the topic of the document. For instance, consider the following:

$$y_i \in \{1, \ldots, C\}.$$

In this case, $y_1 = 1$ might represent "hypertension," $y_2 = 2y$ for "diabetes," and $y_3 = 3$ for "anxiety."

Unlike humans, machines cannot interpret raw text data. To build a machine learning classifier capable of working with textual data, the first step is to convert the text into numerical representations. This process involves transforming each document into a feature vector, where each feature corresponds to a scalar value. These numeric vectors form the foundation for training machine learning algorithms.

One simple and widely-used method for converting text into numerical data is the Bag of Words (BoW) approach. This method treats each document as an unordered collection of words, disregarding grammar, word order, and context. While simplistic, the BoW technique is a practical starting point for many text classification tasks and remains an important foundational concept in natural language processing (NLP).

In clinical decision support (CDS), textual data can come from various sources, such as electronic health records (EHRs), physician notes, or patient-reported outcomes. To illustrate how machine learning can process and analyze textual data, consider the following hypothetical examples from a corpus of medical text.

Item #	Statement
1	Smoking increases the risk of heart disease.
2	Regular exercise improves cardiovascular health.
3	Quitting smoking lowers the risk of heart attack.
4	Managing diabetes involves monitoring blood glucose levels.
5	Lifestyle changes are crucial for diabetes management.
6	Insulin therapy is essential for some patients with diabetes.
7	Hypertension increases the risk of kidney disease.
8	Controlling blood pressure reduces stroke risk.
9	A low-sodium diet helps manage hypertension.
10	Kidney disease requires regular monitoring and follow-up.

FIGURE 2.1 Hypothetical medical text corpus

This collection of text documents, or a corpus, represents clinical insights and recommendations. The BoW method can be applied to analyze this text data and extract valuable features for training a machine learning model. This process involves two steps:

1. Create a Vocabulary

First, compile a list of all unique words in the corpus. This vocabulary serves as the foundation for feature extraction. For example, from the above text, the vocabulary might include words such as the following:

["smoking," "exercise," "diabetes," "heart," "insulin," "stroke," "diet," "monitoring," "risk"].

2. Vectorize the Documents

Next, each document is converted into a feature vector, where each dimension corresponds to a word in the vocabulary. The value of each dimension indicates the presence, absence, or frequency of the word in that document. For instance, consider the following:

- Document 1: "Smoking increases the risk of heart disease." Feature Vector: [1, 0, 0, 1, 0, 0, 0, 0, 1]
  (Here, "smoking," "risk," and "heart" are present, while other words are absent.)

- Document 4: "Managing diabetes involves monitoring blood glucose levels."
  Feature Vector: [0, 0, 1, 0, 0, 0, 0, 1, 0]
  (Here, "diabetes" and "monitoring" are present, while other words are absent.)

By transforming the text into numeric representations, machine learning algorithms can process the data to recognize patterns and make predictions. For example, a model trained on this corpus could identify which topics (e.g., diabetes management or cardiovascular health) are most relevant to a specific patient's clinical scenario. This approach underpins many CDS systems, enabling them to suggest personalized care recommendations or flag potential health risks based on textual information.

Tokenization and Vocabulary Building for Text Processing

In any text analysis task, building a vocabulary is a crucial step that involves identifying all the unique tokens in a corpus. For the given 10-document corpus, the vocabulary is constructed by listing all the distinct words in alphabetical order after applying the following preprocessing steps:

1. Remove Punctuation: Punctuation marks such as periods, commas, and exclamation points are stripped from the text.

2. Convert to Lowercase: All words are converted to lowercase to ensure consistency (e.g., "Movies" and "movies" are treated as the same word).

3. Eliminate Duplicates: Each unique word appears only once in the vocabulary list.

After processing, the resulting vocabulary for the 10-document corpus is as follows:

```
vocabulary = ["abnormal," "acute," "and," "are,"
"assessment," "blood," "chronic," "condition,"
"diabetes," "diagnosis," "discharge," "disease,"
"elevated," "heart," "history," "hypertension,"
"infection," "is," "medication," "monitoring,"
"pain," "patient," "pressure," "recommendation,"
"risk," "sodium," "status," "symptoms,"
"treatment," "vital," "weight"]
```

FIGURE 2.2 Sample vocabulary

Tokenization: Breaking Text into Smaller Units

The process of splitting a document into its smallest indivisible components is called *tokenization*. A *token* typically refers to a single word or subword, depending on the tokenization method. In the example above, the corpus was tokenized into words, making each word the basic unit of analysis. However, tokenization can vary based on the task, language, or computational requirements:

1. word tokenization
   - Words are the most common type of tokens in NLP. For example, the sentence "Movies are fun for everyone." becomes the tokens ["movies," "are," "fun," "for," "everyone"].
   - While this approach is simple and intuitive, it may result in a very large vocabulary, especially for languages with complex morphology (e.g., Finnish or Turkish).
2. subword tokenization
   - In subword tokenization, words are broken into smaller meaningful units or morphemes. For instance, the word "interesting" could be split into ["interest," "-ing"].
   - This approach reduces the size of the vocabulary and is especially useful for handling rare or out-of-vocabulary words. A common technique for subword tokenization is byte-pair encoding (BPE), which iteratively merges the most frequent character pairs to create subword units.
3. character tokenization
   - For some tasks, text is tokenized at the character level, splitting each word into its individual characters (e.g., "movies" becomes ["m," "o," "v," "i," "e," "s"]). While rarely used on its own, this method can be

combined with higher-level tokenization techniques for handling typos, misspellings, or non-standard text inputs.

Why Subword Tokenization Matters

Languages like English, with several million possible surface forms (e.g., "do," "does," "doing," and "did"), and languages with complex morphological structures, such as Finnish or Hungarian, pose challenges for text processing. In Finnish, for instance, a single noun can have 2,000-3,000 different forms to express various grammatical cases and numbers. Storing all these variations as separate tokens in the vocabulary would result in excessive memory usage and computational demands.

Subword tokenization offers a practical solution by doing the following:

- reducing vocabulary size: Instead of storing every inflected form, subwords capture reusable components like roots and suffixes (e.g., "run," "runner," "running" can all share the root "run").
- improving generalization: By breaking down rare or unknown words into subwords, models can better handle words they have not encountered during training. For example, a model trained on "interest" and "-ing" can process "interesting" without needing it explicitly in the training data.
- balancing efficiency: Subword tokenization strikes a balance between the simplicity of word tokenization and the granularity of character tokenization, making it particularly effective in machine learning applications like language modeling and text classification.

Tokens vs. Words

In text processing, tokens are the smallest indivisible units of a document, and the terms "token" and "word" are often used interchangeably. However, the distinction becomes important when subword or character-level tokenization is employed. For example, consider the following:

- In word tokenization, "learning" is a single token.
- In subword tokenization, "learning" might be split into ["learn," "-ing"].
- In character tokenization, it becomes ["l," "e," "a," "r," "n," "i," "n," "g"].

The BoW method, while originally designed for words, can be extended to handle subwords or even characters depending on the application.

Practical Implications in Clinical Text Analysis

In a healthcare context, tokenization plays a critical role in processing clinical narratives, patient-reported outcomes, or research articles. For example, consider the following:

- A sentence like "The patient experienced difficulty breathing and high blood pressure." could be tokenized into words, subwords, or characters depending on the application.
- Subword tokenization could help a model recognize relationships between related terms like "hypertensive" and "hypertension," even if one form is rare in the training data.

By choosing the right tokenization method for the dataset and model, healthcare applications can achieve better accuracy in tasks like clinical note summarization, disease classification, or treatment recommendation systems. This flexibility underscores the importance of tokenization as a foundational step in text preprocessing.

Types of Machine Learning

There are three main types of machine learning: supervised learning, unsupervised learning, and reinforcement learning. Each approach has its own strengths and weaknesses, and the choice of method depends on the nature of the problem and the available data.

2.2.1 Supervised Learning

Supervised learning, according to Hastie et al. (2009), is the most commonly used class of machine learning techniques. It involves training an algorithm on a labeled dataset, where each example consists of an input (or set of inputs) and a known, correct output (label). The model's aim is to learn a mapping function from inputs to outputs so that it can accurately predict labels for new, unseen data.

In a healthcare setting, supervised learning can be particularly useful for the following:

- predicting patient outcomes (e.g., readmission rates or survival probabilities) based on clinical and demographic features
- classifying medical images as normal or abnormal (e.g., detecting cancerous lesions in radiology images)

- identifying risk factors for diseases from structured and unstructured electronic health records

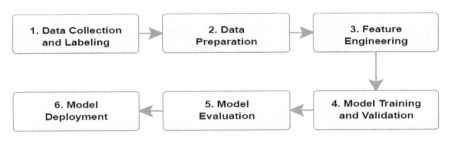

FIGURE 2.3 Important steps in supervised learning

Figure 2.3 outlines the fundamental steps in the supervised learning process. The workflow begins with "1. Data Collection and Labeling," where relevant data is gathered and annotated. Next, "2. Data Preparation" involves cleaning and preprocessing the data for analysis. "3. Feature Engineering" follows, transforming raw data into meaningful features for the model. The prepared data is then used in "4. Model Training and Validation," where algorithms learn from labeled data. The trained model undergoes "5. Model Evaluation" to assess performance metrics and ensure reliability. Finally, in "6. Model Deployment," the optimized model is integrated into real-world applications for predictive tasks.

Data Collection and Labeling

The first step in building any supervised learning model involves identifying and gathering relevant data from sources such as electronic health records (EHRs), clinical trials, or medical imaging databases. Once collected, each instance of data (e.g., a patient record, an imaging study, or a lab result) must be paired with a corresponding label that indicates the desired outcome or ground truth, for instance, a diagnosis, a specific clinical endpoint, or a particular event of interest. Accurate and representative labels are crucial for training the model effectively, as they serve as the reference point the algorithm will learn to predict.

Data Preparation

Once data collection is complete, the raw dataset typically requires extensive cleaning to remove errors, duplicates, and inconsistencies. This phase may also involve handling missing values—either by imputation, removal of incomplete records, or application of domain-driven methods. After cleaning, the data must be transformed into a format amenable to modeling, which may entail normalization (e.g., scaling numeric values), encoding categorical variables (e.g., one-hot encoding for gender or comorbidities), and standardizing data types (e.g., converting dates to numeric features). Proper data preparation helps ensure that the subsequent modeling steps are grounded in reliable and consistent information.

Feature Engineering

Feature engineering focuses on selecting, extracting, and potentially creating relevant variables that capture essential aspects of the clinical problem (Meystre, 2008). For example, in a heart failure readmission model, relevant features might include demographic information, lab test results, comorbidity profiles, and vitals taken at multiple time points. Clinicians and data scientists often collaborate at this stage to identify meaningful transformations—such as trends in laboratory values over time or composite scores of disease severity—that could increase a model's predictive power. Well-designed features often yield more robust and accurate performance than models trained on raw data alone.

Model Training and Validation

To accurately measure performance and avoid overfitting, the dataset is usually split into training, validation, and test sets. The training set is used by the model to learn underlying patterns, while the validation set enables fine-tuning of hyperparameters (e.g., learning rate, tree depth, or number of layers). This iterative tuning process helps identify the best model configuration. By preserving an additional test set that remains untouched until final evaluation, data scientists can gauge how well the chosen model generalizes to new, unseen data. This step is critical for building models that remain reliable when deployed in real-world clinical settings.

Model Evaluation

After training and tuning are complete, the model's performance is assessed against the test set using metrics such as accuracy, F1-score, precision and recall, or the area under the receiver operating characteristic (ROC) curve. These metrics help illustrate how effectively the model can distinguish between classes or predict an outcome of interest. Comparisons might be made across different models (e.g., logistic regression, random forest, or neural networks) or feature subsets to determine which approach yields the highest performance and greatest clinical utility. This stage informs the final selection of the model that will move forward into deployment.

Model Deployment

The final step involves integrating the trained model into a clinical workflow or a decision support system. This deployment can be as simple as exposing a web-based application for real-time predictions, or as complex as embedding the model directly into a hospital's EHR system. Crucially, ongoing monitoring of model performance is necessary to ensure it continues to perform well over time, especially as data distributions and clinical practices evolve. In some cases, model retraining or re-validation may be required to maintain accuracy and relevance. By continuously monitoring these outcomes and involving clinicians in the feedback loop, organizations can ensure that the deployed AI solution remains both accurate and trustworthy.

2.2.2 Common Algorithms in Supervised Learning

Popular supervised learning algorithms, as detailed by James et al. (2021), include the following:

Linear Regression

Linear regression assumes a linear relationship between the input features and a continuous outcome. It is commonly applied in healthcare to predict numeric values such as patient length of stay or lab test results. By fitting a line (or hyperplane in higher dimensions) that minimizes the difference between predicted and actual values, this model provides straightforward interpretability and serves as a foundation for more advanced algorithms.

Logistic Regression

Logistic regression is a classification algorithm that models the probability of a binary outcome—for example, distinguishing between diseased vs. not diseased patients. In healthcare, it is often used to estimate the likelihood of specific health events such as disease onset or readmission. Its probabilistic output and relative interpretability make it a popular choice for many clinical applications.

Decision Trees

Decision trees employ a tree-like model of sequential decisions based on feature values. Each split in the tree corresponds to a threshold on a particular feature, branching off into distinct subgroups. Highly interpretable by design, decision trees are particularly appealing for clinical applications, where transparent decision-making is crucial.

Random Forests

Random forests are an ensemble method that trains multiple decision trees on different subsets of the data and features. Their predictions are then aggregated (often by majority vote) to improve accuracy and reduce overfitting. Because of their robustness, Random forests are widely used for complex datasets, including those with a mixture of categorical and continuous variables.

Support Vector Machines (SVM)

Support Vector Machines seek an optimal hyperplane—or set of hyperplanes in higher-dimensional space—that separates data points of different classes. They can handle nonlinear patterns by using kernel functions, making them highly versatile. However, training SVMs on very large datasets can be computationally expensive and often requires careful hyperparameter tuning.

Applications for Healthcare

Supervised learning has a variety of impactful applications in healthcare. One important use case lies in prognostic modeling, where algorithms help predict a patient's risk of deterioration, readmission, or complications during hospitalization. By identifying those at higher risk, healthcare providers can allocate resources more efficiently and implement proactive interventions that improve patient outcomes.

Another significant application involves diagnostic support, which can assist clinicians in interpreting complex imaging studies such as X-rays, MRIs, and CT scans. By identifying potential anomalies early, these tools reduce the rate of diagnostic errors, a vital benefit in resource-limited settings. In addition, supervised learning contributes to disease risk profiling, identifying individuals at elevated risk for chronic conditions like diabetes or cardiovascular diseases based on their lifestyle and clinical history. This enables targeted preventive care and more personalized treatment pathways.

Supervised learning also plays a central role in treatment response prediction, where models estimate how patients might respond to specific therapies. By utilizing patient data such as genetic markers, comorbidities, and past treatment outcomes, clinicians can tailor interventions to individual needs. This personalized medicine approach ensures that patients and providers select the most effective treatment regimen.

Considerations for Using Supervised Learning in Healthcare

Despite its many benefits, the implementation of supervised learning in healthcare requires careful attention to several important considerations. Data quality and bias are paramount, as a model's reliability depends on the accuracy and representativeness of the data it is trained on. Historical biases embedded in datasets risk perpetuating health disparities unless they are identified and mitigated (Mehrabi, 2021).

In a high-stakes environment like healthcare, interpretability and explainability are equally important. Clinicians and patients alike need clear rationales for how a model arrives at its predictions (Lundberg, 2017). Tools such as LIME or SHAP can shed light on how input features influence specific model outputs, fostering trust and improving clinical adoption.

Additionally, regulatory and ethical concerns must be addressed. Models must comply with healthcare regulations—for instance, HIPAA in the United States—to ensure data privacy and patient safety. Ethical practices around data use and patient consent are vital to maintaining public trust in these technologies.

Effective integration into clinical workflows is another critical factor. Even highly accurate models can fail to deliver tangible benefits if they do not align with day-to-day clinical practices. Training clinical staff, securing user buy-in, and designing intuitive interfaces can help overcome these barriers.

Continuous monitoring and updating are essential, as healthcare data and patient populations evolve over time. Regularly assessing model performance helps detect shifts in accuracy and ensures that any necessary retraining or recalibration is carried out promptly. By staying attuned to these considerations, supervised learning can continue to advance healthcare delivery and patient outcomes.

Predictive modeling and decision support across various domains utilize supervised learning extensively, especially in healthcare. By utilizing labeled datasets and robust machine learning algorithms (Hastie et al., 2009; James et al., 2021), healthcare institutions can gain valuable insights, improve patient care, and streamline clinical decision-making. However, careful attention to data quality, interpretability, ethics, and long-term model maintenance is essential to ensure that these tools truly enhance outcomes for diverse patient populations.

2.2.3 Unsupervised Learning

Unsupervised learning, as described by Hinton and Sejnowski (1999), is a machine learning paradigm in which algorithms are trained on unlabeled data. Unlike supervised methods, where models learn to predict known outputs, unsupervised learning focuses on discovering hidden patterns or structures within the data. This discovery process can include grouping similar observations together, reducing data dimensionality, or detecting anomalies—all without prior knowledge of the "correct" answers.

In healthcare, unsupervised learning offers valuable insights by revealing patterns that may not be immediately visible to clinicians or researchers. For instance, it can be applied to identify subtypes of diseases based on shared patient characteristics, revealing potential variations in treatment responses or disease progression. It can also uncover trends in patient trajectories, helping care teams personalize interventions and track patients over time. Additionally, these techniques can be utilized to detect unusual or fraudulent healthcare claims, supporting more efficient resource allocation and fraud prevention.

Common Techniques in Unsupervised Learning

According to Friedman et al. (2001), there are several established approaches to unsupervised learning:

1. K-means Clustering
    - Partitions data into kkk clusters, each represented by a centroid.
    - Widely used for its simplicity and computational efficiency; however, results depend on the choice of kkk and initial centroid placement.

2. Hierarchical Clustering
    - Builds a tree-like structure (dendrogram) representing nested clusters at multiple levels.

- Useful when the optimal number of clusters is not known in advance; the resulting hierarchy can reveal relationships between sub-clusters.

3. Principal Component Analysis (PCA)
   - A dimensionality reduction method that transforms correlated features into a smaller set of uncorrelated components.
   - Simplifies high-dimensional datasets (e.g., genomics data) to highlight dominant variance patterns, making downstream analyses more manageable.

4. Autoencoders
   - A neural network architecture designed to learn efficient data codings in an unsupervised manner.
   - Particularly useful for anomaly detection, denoising signals (e.g., ECG waveforms), or extracting compressed representations for further analysis.

Applications for Healthcare

Unsupervised learning has several impactful applications in healthcare. One notable example is identifying subtypes of diseases, where clustering algorithms group patients with similar clinical or genetic profiles (Liao et al., 2016). By uncovering previously unrecognized disease subtypes, these methods can lead to more targeted interventions. For instance, clustering might reveal distinct metabolic profiles among diabetes patients, prompting individualized treatment strategies.

Another significant application involves discovering patterns in patient trajectories. Time-series data, such as vital signs, lab results, and treatment timelines, can be clustered or reduced in dimensionality to pinpoint common care pathways or disease progressions. These insights enable clinicians to tailor follow-up schedules and optimize long-term disease management plans.

Unsupervised learning also aids in fraud and anomaly detection by identifying unusual behavior in healthcare claims or billing data. After learning what constitutes typical claims, the algorithm highlights outliers for further investigation. This automated approach helps organizations detect fraud or abuse more quickly, potentially saving costs and preserving resources for legitimate care.

Clinical pathway optimization benefits from techniques like dimensionality reduction, which can simplify large clinical datasets and reveal potential inefficiencies in patient workflows. Cluster analyses may further group patients with similar care needs, guiding hospitals to allocate resources more effectively and enhance overall patient outcomes.

Considerations for Using Unsupervised Learning in Healthcare

Despite its promise, unsupervised learning also presents unique challenges. Interpretability is one key hurdle: these models can generate complex clusters or patterns that are not inherently intuitive. Collaborating with clinical experts and using visualization tools such as cluster heatmaps can help translate findings into actionable clinical insights.

The quality and potential biases in healthcare data are equally critical. Missing or inconsistent data, as well as non-representative samples, can distort the patterns that unsupervised algorithms detect. Addressing these issues through careful data cleaning, imputation methods, and bias mitigation is essential to ensure reliable conclusions.

Scalability and computational constraints can also pose problems, as certain algorithms (e.g., hierarchical clustering) demand substantial computational resources when dealing with very large datasets. Strategies such as efficient sampling, parallelization, or distributed computing may be necessary in big data environments.

Reliable clinical validation for any novel insights derived from unsupervised analysis is essential. Patterns or clusters suggested by an algorithm require assessment by clinicians and domain experts to confirm their relevance and consistency with real-world patient care.

Finally, ethical and regulatory concerns must be considered carefully. Even though unsupervised learning does not rely on labels, it still involves patient data that must comply with privacy regulations like HIPAA. Sensitive patient characteristics or inferred clusters should be carefully managed, with robust de-identification measures and secure data storage in place.

Unsupervised learning provides powerful tools for uncovering hidden patterns, structures, and relationships in unlabeled healthcare data (Hinton & Sejnowski, 1999; Friedman et al., 2001). From identifying novel disease subtypes to optimizing clinical workflows, these techniques offer new insights that may not be immediately apparent to human observers. Nonetheless, realizing the full potential of unsupervised methods requires careful attention to data quality, interpretability, and ongoing clinical collaboration. By integrating domain expertise with robust data science practices, healthcare organizations can utilize unsupervised learning to encourage innovation and improve patient outcomes.

2.2.4 Reinforcement Learning

Reinforcement learning, as described by Sutton and Barto (2018), is a branch of machine learning in which an agent learns to make decisions by actively interacting with its environment. Each action taken by the agent results in a reward or penalty, guiding the agent to optimize its cumulative reward over time. This learning paradigm differs from supervised and unsupervised approaches by emphasizing trial-and-error methods rather than labeled data or hidden pattern discovery.

In healthcare, reinforcement learning has shown notable potential across several domains. One application involves optimizing treatment plans by dynamically adjusting interventions based on a patient's responses, aiming to maximize therapeutic benefits while minimizing risks or side effects. Reinforcement learning can also improve hospital resource allocation by learning to optimally schedule staffing, manage bed capacity, or direct patients to the most suitable care setting. Additionally, it can support personalized health interventions, tailoring recommendations to individual preferences and behaviors in areas such as medication adherence, lifestyle changes, and chronic disease management.

Popular reinforcement learning algorithms, such as Q-learning, SARSA, and policy gradient methods (Mnih et al., 2015), enable agents to adapt and refine their decision-making strategies through continuous feedback from the healthcare environment. For instance, policy gradient methods can handle complex decision spaces by learning a direct mapping from states to actions, while Q-learning and SARSA are often used in more discrete state-action settings. Through careful formulation of rewards and states that reflect real-world healthcare objectives, these algorithms can power intelligent systems that learn to optimize outcomes for patients, clinicians, and healthcare institutions alike.

2.2.5 Clinical Applications of Machine Learning

Machine learning is already being used in a variety of clinical applications to improve patient care, optimize processes, and support decision-making. Some examples include the following:

- predicting hospital readmissions and identifying high-risk patients for interventions (Rajkomar et al., 2018)
- detecting diseases such as cancer, diabetic retinopathy, and Alzheimer's from medical images (Esteva et al., 2017; Gulshan et al., 2016; Ding et al., 2018)
- identifying adverse drug events and potential drug interactions from electronic health records (Tatonetti et al., 2012)
- optimizing hospital staffing and resource allocation based on patient demand forecasting (Kutafina et al., 2019)

As machine learning techniques continue to advance and more healthcare data becomes available, the potential applications of these technologies will likely expand. However, it is important to recognize the limitations and challenges of machine learning in healthcare, such as the need for large, high-quality datasets, the risk of bias and errors, and the importance of human oversight and interpretation (Char et al., 2018).

Practical Tip: When considering the use of machine learning in a clinical setting, it is essential to carefully evaluate the quality and representativeness of the data, the appropriateness of the chosen algorithm, and the potential impact on patient care. Engaging with data scientists, clinical experts, and other stakeholders can help ensure that machine learning is applied responsibly and effectively.

Code Tip: Below is a simple demonstration of how one might predict hospital readmissions using a random forest classifier on synthetic data. We first generate a mock dataset of patient information (such as age, comorbidities, and prior admissions) and a label indicating whether the patient was readmitted within 30 days. We then train the model, evaluate its performance, and finally identify high-risk patients based on predicted probabilities. In a real-world scenario, you would replace the synthetic data creation with actual patient data (appropriately secured and de-identified) and potentially refine the model with further features and hyperparameter tuning.

Simulating a Synthetic Dataset for Hospital Readmissions

```python
# Import necessary libraries
import numpy as np
import pandas as pd
from sklearn.model_selection import train_test_split
from sklearn.ensemble import RandomForestClassifier
from sklearn.metrics import classification_report, confusion_matrix

# Simulate patient data
np.random.seed(42)  # Ensures reproducible random numbers
num_samples = 1000  # Number of samples in the dataset

# Create synthetic features
age = np.random.randint(18, 90, size=num_samples)  # Patient age between 18 and 90
num_comorbidities = np.random.randint(0, 5, size=num_samples)  # Number of comorbidities (0-4)
prior_admissions = np.random.randint(0, 10, size=num_samples)  # Number of prior hospital admissions
days_in_hospital = np.random.randint(1, 15, size=num_samples)  # Length of hospital stay (1-14 days)

# Generate readmission labels using a risk scoring mechanism
base_risk_score = (
    (num_comorbidities * 1.5) +
```

```
        (prior_admissions * 0.7) +
        (days_in_hospital * 1.2)
)
prob_of_readmission = 1 / (1 + np.exp(-0.2 * (base_
risk_score - 8)))   # Logistic function
readmitted = (prob_of_readmission > 0.5).astype(int)
# Patients with probability > 0.5 are labeled as
readmitted

# Create a DataFrame to hold the dataset
data = pd.DataFrame({
    'age': age,
    'num_comorbidities': num_comorbidities,
    'prior_admissions': prior_admissions,
    'days_in_hospital': days_in_hospital,
    'readmitted': readmitted
})
```

Explanation

- This section creates a synthetic dataset with features relevant to hospital readmissions.
- Features (age, num_comorbidities, prior_admissions, and days_in_hospital) are generated using random values to simulate patient data.
- Readmission labels (readmitted) are generated using a logistic function based on a calculated risk score, ensuring patients with higher risk factors are more likely to be labeled as readmitted.
- The dataset is stored in a Pandas data frame, which is easy to manipulate and analyze.

```
X = data.drop('readmitted', axis=1)   # Features
y = data['readmitted']   # Target variable

X_train, X_test, y_train, y_test = train_test_split(
    X, y, test_size=0.2, random_state=42
)
```

Explanation

- X: contains the input features used for prediction (all columns except readmitted)
- y: contains the target variable (readmitted), indicating whether the patient was readmitted
- train_test_split: splits the dataset into
  - training set (80%) for model training
  - testing set (20%) for evaluating the model's performance
- random_state=42 ensures the split is reproducible

```
clf = RandomForestClassifier(n_estimators=100, random_state=42)   # Initialize the model
clf.fit(X_train, y_train)   # Train the model on the training data
```

Explanation

- RandomForestClassifier
  - a machine learning model that combines multiple decision trees (ensemble)
  - Each tree votes for a class, and the majority vote determines the final prediction.
- n_estimators=100: specifies 100 decision trees in the forest
- fit: trains the model using the training data (X_train, y_train)

```
y_pred = clf.predict(X_test)   # Predict labels for the test set

print("Classification Report:")

print(classification_report(y_test, y_pred))   # Detailed performance metrics

print("Confusion Matrix:")

print(confusion_matrix(y_test, y_pred))   # Tabular summary of prediction results
```

Explanation

- predict: uses the trained model to predict labels for the test set (X_test)
- classification_report
    - provides detailed performance metrics, including the following:
        - precision: how many predicted positives are actual positives
        - recall: how many actual positives are correctly identified
        - F1-score: harmonic mean of precision and recall
        - support: number of true instances for each class
- confusion_matrix
    - provides a matrix of True Positives (TP), True Negatives (TN), False Positives (FP), and False Negatives (FN)

```
y_prob = clf.predict_proba(X_test)[:, 1]
# Probability of readmission for each patient
risk_threshold = 0.7
# Define a threshold for high risk
```

```
high_risk_indices = np.where(y_prob > 
risk_threshold)[0]
# Find indices of high-risk-risk patients

print(f"\nNumber of high-risk-risk patients (prob > 
{risk_threshold}): {len(high_risk_indices)}")

# Extract high-risk patients' data
high_risk_patients = X_test.iloc[high_risk_indices].
copy()
high_risk_patients['predicted_readmission_probability'] 
= y_prob[high_risk_indices]

print("\nSample of high-risk patients:")
print(high_risk_patients.head())
```

Explanation

- predict_proba
  - returns the probability of each class (readmitted = 0 or 1) for each patient
  - [:, 1] extracts the probability of the positive class (readmitted=1)
- risk_threshold=0.7
  - Patients with a predicted probability > 0.7 are classified as "high risk."
- np.where(y_prob > risk_threshold)
  - identifies indices of patients whose predicted probability exceeds the threshold
- high_risk_patients
  - extracts data records of high-risk patients from X_test
  - adds a new column (predicted_readmission_probability) to display their predicted probabilities

How This All Works:

Data Simulation

1. We create a synthetic dataset that includes basic patient features (e.g., age, comorbidities, prior admissions). The variable readmitted is artificially determined based on a logistic function of those features.
2. Train/Test Split
3. We split our dataset into training (80%) and testing (20%) sets to prevent overfitting and assess how well the model generalizes.
4. Model Training
5. We train a random forest classifier on the training data. In practice, you would also want to tune hyperparameters (e.g., using GridSearchCV or RandomizedSearchCV) to achieve optimal performance.

Evaluation

- We predict on the test set and generate a classification report (accuracy, precision, recall, and F1-score) and a confusion matrix.

Identifying High-Risk Patients

- We retrieve the model's predicted probabilities (predict_proba) to rank patients by their likelihood of readmission. Those exceeding an arbitrary threshold (e.g., 70%) could be flagged for additional interventions (e.g., scheduling more frequent follow-up visits).

Important Note

- This code is strictly for demonstration and uses synthetic data. You would replace the data generation step with real patient data (secured and de-identified per HIPAA/GDPR, if applicable).
- In production, you would include robust data preprocessing steps and model explanation or interpretability techniques (e.g., SHAP values) to ensure the model's output can be used responsibly in clinical practice.

CHAPTER 3

Neural Networks and Deep Learning

3.1 NEURAL NETWORKS

Neural networks, a class of machine learning models inspired by the structure and function of the human brain, have revolutionized AI in recent years, particularly in the field of healthcare. By connecting simple processing units in complex architectures, neural networks can learn to recognize patterns, make decisions, and solve problems in ways that often surpass human capabilities. *Deep learning*, a subfield of neural networks, has achieved remarkable breakthroughs in areas such as medical image analysis, electronic health record (EHR) processing, and clinical decision support.

In healthcare, deep learning is being applied to analyze medical images, predict disease progression, and personalize treatment plans, among other applications. For instance, convolutional neural networks (CNNs) have been used to detect tumors, lesions, and abnormalities in X-rays, CT scans, and MRIs, while recurrent neural networks (RNNs) have been employed to predict patient outcomes and generate medical reports from EHR data.

This section explores the fundamentals of neural networks and deep learning, focusing on their applications in healthcare. We discuss the structure of neural networks, common deep learning architectures, and their potential to transform clinical decision-making (Singhal et al., 2020). Additionally, we address the challenges and limitations of deep learning in healthcare, such as interpretability, bias, and fairness, and provide practical tips for developing and deploying these models in clinical settings.

3.2 THE STRUCTURE OF NEURAL NETWORKS IN HEALTHCARE APPLICATIONS

Neural networks consist of interconnected nodes, or *neurons*, organized into layers that process and transmit information. Each neuron receives input from the previous layer, performs a computation, and passes its output to the next layer. By adjusting the strengths of the connections between neurons, known as *weights*, the network can learn to map inputs to outputs and perform complex tasks in healthcare settings (Goodfellow et al., 2016).

A typical neural network used in healthcare applications consists of three main types of layers: input, hidden, and output. The input layer receives the initial data, such as patient features or medical images. The hidden layers transform and extract meaningful representations of the input data. The output layer produces the final predictions or classifications based on the learned representations (LeCun et al., 2015).

For example, in a CNN designed to classify skin lesions from dermoscopic images, the input layer would receive the pixel values of the image, the hidden layers would learn to detect relevant features such as color, texture, and shape, and the output layer would predict the probability of the lesion being benign or malignant (Esteva et al., 2017).

The computation performed by each neuron involves two main steps: aggregation and activation. In the aggregation step, the neuron computes a weighted sum of its inputs, where each input is multiplied by the corresponding connection weight. This weighted sum represents the neuron's total input signal. In the activation step, the neuron applies a nonlinear activation function to the weighted sum, which introduces nonlinearity into the network and allows it to learn complex relationships between inputs and outputs (Nair & Hinton, 2010).

Some common activation functions used in healthcare neural networks include the following:

- Sigmoid: A smooth, S-shaped curve that *squashes* (compresses) the input to a value between 0 and 1, often used in the output layer for binary classification tasks, such as predicting the presence or absence of a disease.

- Rectified Linear Unit (ReLU): A piecewise linear function that outputs the input if it is positive, and 0 otherwise, commonly used in the hidden layers to introduce sparsity and reduce computational cost.

The choice of activation function depends on the specific healthcare problem and the desired properties of the network, such as the range of the outputs and the sparsity of the activations (Glorot et al., 2011).

During training, the neural network learns to map inputs to outputs by adjusting its weights based on the difference between the predicted and actual outputs, using a process called *backpropagation* (Rumelhart et al., 1986). By iteratively presenting examples to the network and updating the weights, the neural network can learn to approximate the underlying function that relates inputs to outputs, such as mapping patient features to disease risk scores or medical images to diagnostic labels.

Practical Tip: When designing a neural network for a healthcare application, it is important to consider the nature of the input data, the desired output, and the available computational resources. Starting with a simple architecture and gradually increasing the complexity can help identify the optimal balance between performance and efficiency. Regularization techniques such as L1/L2 regularization, dropout, and early stopping can help prevent overfitting and improve the generalization of the network (Goodfellow et al., 2016).

3.3 DEEP LEARNING ARCHITECTURES IN HEALTHCARE

Deep learning is a subfield of machine learning that focuses on neural networks with many layers, known as *deep neural networks* (Goodfellow et al., 2016). These architectures can learn hierarchical representations of data, where each layer captures increasingly abstract and complex features. Deep learning has achieved remarkable success in a wide range of healthcare domains, including medical image analysis, EHR processing, and clinical decision support (Esteva et al., 2019).

Some common deep learning architectures used in healthcare include the following:

1. Convolutional Neural Networks (CNNs): CNNs are designed to process grid-like data, such as medical images or time series, by learning local patterns and spatial hierarchies (LeCun et al., 1998). They consist of convolutional layers that apply learned filters to the input, pooling layers that downsample the feature maps, and fully connected layers that perform

the final prediction or classification. CNNs have been widely used in medical image analysis tasks, such as

- detecting lung nodules, breast lesions, and brain tumors from CT, mammography, and MRI scans (Ardila et al., 2019; McKinney et al., 2020; Kamnitsas et al., 2017)
- classifying skin lesions, retinal images, and histopathology slides for early detection of skin cancer, diabetic retinopathy, and other diseases (Gulshan et al., 2016; Litjens et al., 2016)
- segmenting organs, tissues, and anatomical structures in medical images for surgical planning and radiation therapy (Ronneberger et al., 2015)

2. Recurrent Neural Networks (RNNs): RNNs are designed to process sequential data, such as time series or EHR data, by maintaining a hidden state that captures information from previous time steps (Lipton et al., 2015). They have feedback connections that allow them to model temporal dependencies and learn long-term patterns. Some popular RNN architectures used in healthcare include Long Short-Term Memory (LSTM) and Gated Recurrent Units (GRUs), which use gating mechanisms to control the flow of information and mitigate the vanishing gradient problem (Hochreiter & Schmidhuber, 1997; Cho et al., 2014). RNNs have been applied to various healthcare tasks, such as

- predicting patient outcomes, such as mortality, readmission, or disease progression, from EHR data (Rajkomar et al., 2018; Tomašev et al., 2019)
- generating medical reports, such as radiology findings or discharge summaries, from images or structured data (Jing et al., 2017; Liu et al., 2019)
- analyzing time series data, such as ECG or EEG signals, for anomaly detection or disease diagnosis (Chauhan & Vig, 2015)

3. Generative Adversarial Networks (GANs): GANs are a class of deep learning models that learn to generate realistic samples by pitting two networks against each other: a generator that creates new examples and a discriminator that tries to distinguish real from generated examples (Goodfellow et al., 2014). By training the networks in a minimax game, GANs can learn to capture the underlying distribution of the data and generate diverse

and high-quality samples. GANs have been used in healthcare for tasks such as

- synthesizing realistic medical images, such as CT scans or mammograms, for data augmentation or anonymization (Frid-Adar et al., 2018)
- generating synthetic patient data, such as EHR records or time series, for research or privacy protection (Choi et al., 2017)
- translating between different imaging modalities, such as from MRI to CT or from low-dose to high-dose scans (Wolterink et al., 2017)

While deep learning has shown impressive results in many healthcare applications, there are also challenges and limitations to consider. Deep neural networks often require large amounts of labeled data for training, which can be difficult or expensive to obtain in medical domains. They can also be sensitive to data quality issues, such as noise, artifacts, or missing values, which are common in real-world healthcare settings. Deep learning models can be complex and opaque, making it difficult to interpret their predictions or explain their reasoning to clinicians and patients (Miotto et al., 2018).

Practical Tip: When applying deep learning to healthcare problems, it is important to carefully validate the models on diverse and representative datasets and to engage with domain experts to assess the clinical relevance and plausibility of the results. Using techniques such as transfer learning, data augmentation, and unsupervised pre-training can help reduce the need for large labeled datasets and improve the robustness of the models (Raghu et al., 2019). Incorporating attention mechanisms, feature visualization, and model-agnostic explanations can also help improve the interpretability and trustworthiness of deep learning in healthcare (Tonekaboni et al., 2019).

3.4 CLINICAL IMPLICATIONS OF DEEP LEARNING

Deep learning has the potential to revolutionize healthcare by enabling more accurate, efficient, and personalized patient care. By utilizing the ability of deep neural networks to learn complex patterns and representations from large-scale medical data, deep learning can assist clinicians with various healthcare tasks, from diagnosis and prognosis to treatment planning and monitoring (Topol, 2019).

One of the most promising applications of deep learning in healthcare is medical image analysis. Deep learning models, particularly CNNs, have achieved remarkable performance in detecting and classifying various diseases and abnormalities from medical images, often surpassing human experts (Litjens et al., 2017). For example, deep learning algorithms have been developed to

- detect lung nodules, breast lesions, and brain tumors from CT, mammography, and MRI scans (Ardila et al., 2019; McKinney et al., 2020; Kamnitsas et al., 2017)
- identify diabetic retinopathy, glaucoma, and age-related macular degeneration from retinal fundus images (Gulshan et al., 2016; Li et al., 2018)
- classify skin lesions, including melanoma, from dermoscopic images (Esteva et al., 2017)
- segment organs, tissues, and anatomical structures from medical images for surgical planning and radiation therapy (Ronneberger et al., 2015)

By automating the analysis of medical images, deep learning can help reduce the workload of radiologists and pathologists, improve the accuracy and consistency of diagnoses, and enable earlier detection of diseases when they are more treatable.

Another important application of deep learning in healthcare is the analysis of EHR data and other clinical data. Deep learning models, such as RNNs and transformer architectures, can process sequential and unstructured data in EHRs, including clinical notes, lab results, and medication histories, to extract meaningful insights and make predictions (Shickel et al., 2018). For example, deep learning has been used to

- predict patient outcomes, such as mortality, readmission, and disease progression, based on EHR data (Rajkomar et al., 2018; Tomašev et al., 2019)
- identify patients at risk of developing chronic diseases, such as diabetes, heart failure, and Alzheimer's, for early intervention and prevention (Choi et al., 2016; Makino et al., 2019)
- discover novel drug-drug interactions and adverse drug events from EHR data (Zitnik et al., 2018)
- generate natural language summaries of patient records for clinical decision support and patient communication (Liu et al., 2019)

By utilizing the vast amounts of data in EHRs, deep learning can help clinicians make more informed decisions, optimize treatment plans, and improve patient outcomes.

Despite the promising potential of deep learning in healthcare, there are also significant challenges and limitations that need to be addressed. One major challenge is the interpretability and transparency of deep learning models (Gilpin et al., 2018). Unlike traditional statistical models, deep neural networks are often seen as "black boxes" that make predictions based on complex and opaque patterns learned from data. This lack of interpretability can hinder the trust and adoption of deep learning in clinical practice, where clinicians need to understand and explain the reasoning behind their decisions (Ahmad et al., 2018).

Another challenge with deep learning in healthcare is the potential for bias and fairness issues (Char et al., 2018). Deep learning models can inherit and amplify biases present in the training data, such as underrepresentation of certain demographic groups or systematic differences in data quality and completeness. This can lead to disparities in model performance and outcomes across different subpopulations, raising ethical concerns about the equitable deployment of deep learning in healthcare (Rajkomar et al., 2018).

Practical Tip: When developing and deploying deep learning models in healthcare, it is crucial to engage with clinicians, patients, and other stakeholders to ensure that the models are clinically relevant, ethically sound, and aligned with the needs and values of the healthcare system. Interdisciplinary collaboration between machine learning researchers, clinicians, ethicists, and policymakers can help address the technical, clinical, and societal challenges of deep learning in healthcare and realize its full potential for improving patient care (Doshi-Velez & Kim, 2017; Char et al., 2020).

3.5 PROGRAMMING A SIMPLE NEURAL NETWORK FOR HEALTHCARE

To illustrate the concepts discussed in this chapter, let's implement a simple neural network for a healthcare task using PyTorch, a popular deep learning framework. We create a model to predict the likelihood of a patient having diabetes based on their clinical features.

First, we import the necessary libraries and define the neural network architecture:

```
import torch
import torch.nn as nn
import numpy as np
from sklearn.preprocessing import StandardScaler
```

Next, we define the neural network architecture. For diabetes prediction, we create a model that goes beyond the basic feedforward network by incorporating best practices for clinical applications.

```
class DiabetesPredictionModel(nn.Module):
    def __init__(self, input_dim, hidden_dim, output_dim, dropout_rate=0.2):
        super(DiabetesPredictionModel, self).__init__()
        self.fc1 = nn.Linear(input_dim, hidden_dim)
        self.bn1 = nn.BatchNorm1d(hidden_dim)
        self.relu = nn.ReLU()
        self.dropout = nn.Dropout(dropout_rate)
        self.fc2 = nn.Linear(hidden_dim, output_dim)
        self.sigmoid = nn.Sigmoid()

        # Initialize weights using He/Kaiming initialization
        nn.init.kaiming_normal_(self.fc1.weight)
        nn.init.kaiming_normal_(self.fc2.weight)
```

The model architecture includes several important components for clinical applications. Batch normalization (nn.BatchNorm1d) helps stabilize learning and allows faster training with higher learning rates. We add dropout as a regularization technique to prevent overfitting, which is particularly important in medical models where training data may be limited. The weights are initialized using He (Kaiming) initialization, which is well-suited for ReLU activations and helps prevent vanishing gradients.

Now we define the forward pass through our neural network:

```
def forward(self, x):
    x = self.fc1(x)
    x = self.bn1(x)
    x = self.relu(x)
    x = self.dropout(x)
    x = self.fc2(x)
    x = self.sigmoid(x)
    return x
```

This model first processes the input through a fully connected layer, applies batch normalization, ReLU activation, and dropout for regularization. The final layer applies a sigmoid activation to output a probability between 0 and 1, representing the likelihood of diabetes. The input features typically include clinical parameters such as age, BMI, blood pressure, and glucose levels.

Data Preparation and Model Initialization

Proper data preparation is critical for clinical decision support systems. For our diabetes prediction model, we need to normalize the features and split the data into training and validation sets.

```
# Assume we have a dataset of patient features (X) and labels (y)
X = torch.tensor([[0.2, 0.3, 0.1, 0.4], [0.5, 0.1, 0.6, 0.3], ...], dtype=torch.float32)
y = torch.tensor([[1], [0], ...], dtype=torch.float32)

# Normalize the features
scaler = StandardScaler()
X_normalized = torch.tensor(scaler.fit_transform(X), dtype=torch.float32)
```

Normalization is essential for neural networks to perform optimally. Clinical features often have different scales (e.g., age in years vs. glucose in mg/dL), and normalization ensures that each feature contributes proportionally to the model's learning process.

Next, we split our data into training and validation sets, which is crucial for monitoring model performance and preventing overfitting:

```
# Split data into training and validation sets
train_size = int(0.8 * len(X_normalized))
X_train, X_val = X_normalized[:train_size], X_normalized[train_size:]
y_train, y_val = y[:train_size], y[train_size:]
```

With our data prepared, we initialize the model with the appropriate dimensions:

```
input_dim = X_train.shape[1]
hidden_dim = 128
output_dim = 1

model = DiabetesPredictionModel(input_dim, hidden_dim, output_dim)
```

Loss Function, Optimizer, and Evaluation Metrics

For binary classification problems like diabetes prediction, Binary Cross Entropy (BCE) is the appropriate loss function. We use the Adam optimizer, which adapts the learning rate during training for better convergence.

```
criterion = nn.BCELoss()
optimizer = torch.optim.Adam(model.parameters(), lr=0.001)
```

In clinical settings, we need more than just accuracy to evaluate model performance. Metrics like sensitivity (recall) and specificity are critical for assessing the real-world utility of the model:

```
def calculate_metrics(model, X, y):
    with torch.no_grad():
        outputs = model(X)
        predicted = (outputs > 0.5).float()
        accuracy = (predicted == y).float().mean()
        sensitivity = ((predicted == 1) & (y == 1)).sum() / max((y == 1).sum(), 1)
        specificity = ((predicted == 0) & (y == 0)).sum() / max((y == 0).sum(), 1)
    return accuracy.item(), sensitivity.item(), specificity.item()
```

Sensitivity measures the model's ability to correctly identify patients with diabetes (true positive rate), while *specificity* measures its ability to correctly identify patients without diabetes (true negative rate). Both metrics are crucial in clinical decision support to balance the risks of false positives and false negatives.

Model Training with Early Stopping

When training models for clinical applications, it is important to implement early stopping to prevent overfitting to the training data, which could lead to poor generalization in real-world scenarios.

```
num_epochs = 100
batch_size = 32
best_val_loss = float('inf')
patience = 10
patience_counter = 0
```

We train the model in batches and monitor validation performance after each epoch:

```
for epoch in range(num_epochs):
    model.train()
    for i in range(0, len(X_train), batch_size):
        batch_X = X_train[i:i+batch_size]
        batch_y = y_train[i:i+batch_size]

        optimizer.zero_grad()
        outputs = model(batch_X)
        loss = criterion(outputs, batch_y)
        loss.backward()
        optimizer.step()

    # Validation phase
    model.eval()
    with torch.no_grad():
        val_outputs = model(X_val)
        val_loss = criterion(val_outputs, y_val)
        accuracy, sensitivity, specificity = calculate_metrics(model, X_val, y_val)

    print(f"Epoch {epoch+1}/{num_epochs}, Val Loss: {val_loss:.4f},"
          f"Accuracy: {accuracy:.4f}, Sensitivity: {sensitivity:.4f},"
          f"Specificity: {specificity:.4f}")
```

The model.train() and model.eval() calls are important as they toggle the behavior of certain layers like BatchNorm and Dropout between training and evaluation modes.

Now we implement early stopping to save the best model and stop training when performance on the validation set stops improving:

```python
# Early stopping
    if val_loss < best_val_loss:
        best_val_loss = val_loss
        patience_counter = 0
        torch.save(model.state_dict(), 'best_diabetes_model.pt')
    else:
        patience_counter += 1
        if patience_counter >= patience:
            print(f"Early stopping at epoch {epoch+1}")
            break

# Load the best model for inference
model.load_state_dict(torch.load('best_diabetes_model.pt'))
```

Early stopping is crucial in clinical applications where we want to ensure the model generalizes well to new patients rather than memorizing patterns in the training data.

Model Inference and Interpretability

After training, we can use our model to predict the probability of diabetes for new patients:

```
# Ensure new patient data is normalized using the same scaler
new_patient_raw = torch.tensor([[0.3, 0.2, 0.4, 0.5]])
new_patient = torch.tensor(scaler.transform(new_patient_raw), dtype=torch.float32)

model.eval()
predicted_prob = model(new_patient)
print(f"The predicted probability of diabetes is: {predicted_prob.item():.2f}")
```

For clinical decision support systems, model interpretability is not just a nice-to-have feature but often a regulatory requirement. Healthcare professionals need to understand why a model made a particular prediction to trust and appropriately incorporate it into clinical workflows.

```
import shap

# Create a background dataset for SHAP
background = X_train[:100]

# Create an explainer
explainer = shap.DeepExplainer(model, background)

# Calculate SHAP values for the new patient
shap_values = explainer.shap_values(new_patient)

# Get feature names (assuming we have them)
feature_names = ["Age", "BMI", "Blood Pressure", "Glucose"]
```

```
# Print feature importance
for i, feature in enumerate(feature_names):
    print(f"{feature}: Contribution to prediction = {shap_values[0][0][i]:.4f}")
```

The SHAP (SHapley Additive exPlanations) framework provides a unified approach to explaining model outputs. By calculating the contribution of each feature to the prediction, clinicians can better understand which factors are driving the model's decisions, helping to build trust and potentially revealing clinically relevant insights.

This diabetes prediction model demonstrates how modern machine learning techniques can be applied to clinical decision support systems while ensuring the model is accurate, robust, and interpretable for healthcare applications.

3.6 FUTURE DIRECTIONS

Deep learning has become incredibly useful in healthcare, demonstrating its potential to transform various aspects of patient care, from diagnosis and prognosis to treatment planning and drug discovery. As the field continues to evolve, there are several exciting future directions and opportunities for applying deep learning.

One promising direction is the integration of deep learning with other advanced technologies, such as blockchain, edge computing, and 5G networks, to enable secure, decentralized, and real-time processing of medical data. For example, federated learning frameworks can allow deep learning models to be trained on distributed datasets across multiple institutions without sharing sensitive patient data, enabling large-scale collaborative research while preserving privacy (Rieke et al., 2020).

Another important future direction is the development of explainable and trustworthy deep learning models for healthcare. As deep learning becomes more widely adopted in clinical practice, there is a growing need for models that can provide clear and interpretable explanations for their predictions, allowing clinicians to understand and validate the reasoning behind the model's decisions (Tonekaboni et al., 2019). Techniques such as attention mechanisms, feature visualization, and model-agnostic explanations are being explored to improve the transparency and accountability of deep learning in healthcare (Ahmad et al., 2018).

There is no doubt that a need exists for more diverse and representative datasets to train and evaluate deep learning models in healthcare. Current datasets often suffer from biases and limitations, such as underrepresentation of certain demographic groups, lack of diversity in imaging modalities and protocols, and inconsistencies in data quality and labeling (Willemink et al., 2020). Efforts to create large-scale, high-quality, and inclusive datasets, such as the UK Biobank and the NIH All of Us Research Program, are crucial for developing robust and equitable deep learning models that can generalize across different populations and settings.

Finally, the successful translation of deep learning from research to clinical practice will require close collaboration and partnerships between machine learning researchers, healthcare providers, industry stakeholders, and regulatory agencies. Interdisciplinary teams and initiatives, such as the NIH Bridge2AI program and the MICCAI Educational Initiative, are working to foster dialogue, knowledge sharing, and best practices for the development and deployment of deep learning in healthcare (Clark et al., 2024). Establishing clear guidelines, standards, and evaluation frameworks for deep learning in healthcare will be essential to ensure its safety, efficacy, and ethical use in patient care.

Deep learning has emerged as a powerful and transformative technology, with the potential to revolutionize various aspects of patient care. By utilizing the ability of deep neural networks to learn complex patterns and representations from large-scale medical data, deep learning can assist clinicians in tasks ranging from diagnosis and prognosis to treatment planning and drug discovery. However, realizing its full potential in healthcare will require addressing significant challenges and limitations, such as interpretability, fairness, and generalizability, through interdisciplinary collaboration and partnerships. As these capabilities continue to flourish, there are exciting opportunities for deep learning to contribute to more accurate, efficient, and personalized patient care, ultimately improving health outcomes and quality of life for individuals and populations worldwide.

CHAPTER 4

Natural Language Processing

Natural language processing (NLP) is a foundational concept in AI that enables computers to understand, interpret, and generate human language. In healthcare, NLP is being used to extract valuable insights from unstructured data sources, such as clinical notes, patient narratives, and medical literature. By automating the analysis of text data, NLP can help healthcare providers make more informed decisions, improve the quality and efficiency of care, and advance medical research. In this section, we explore the important applications of NLP in healthcare, including text mining, sentiment analysis, and clinical documentation. We also discuss the challenges and opportunities associated with applying NLP to the complex and diverse language of medicine.

4.1 THE TRANSFORMER ARCHITECTURE

The transformer has revolutionized natural language processing in recent years, enabling breakthroughs in machine translation, text generation, and language understanding. It overcomes the important limitations of recurrent neural networks (RNNs) by introducing the self-attention mechanism, which allows the model to process sequences in parallel rather than one token at a time.

There are three main types of transformer architectures:

- *Encoder-decoder:* Originally proposed for machine translation, this architecture consists of an encoder that processes the input sequence and a decoder that generates the output sequence. The encoder and decoder are connected via an attention mechanism.

- *Encoder-only*: Typically used for tasks like text classification and named entity recognition, this architecture consists of just an encoder stack. The encoder's output is pooled and passed to a classification head.

- *Decoder-only*: Most commonly used for language modeling and text generation, this architecture consists of just a decoder stack. Each token attends to all previous tokens to predict the next one. This is the architecture we focus on.

The innovations of the transformer are self-attention and positional encoding. Self-attention allows each token to attend to all other tokens in the sequence, enabling the model to capture long-range dependencies. Positional encoding injects information about the position of each token, since the model has no inherent notion of token order.

4.1.1 Decoder Architecture

The decoder-only transformer is designed to process a sequence of tokens and generate an output sequence, where each output token is conditioned on the previous tokens. This is known as *autoregressive modeling* and is particularly well-suited to tasks like language modeling, where the goal is to predict the next token given the previous tokens.

The architecture consists of a stack of identical decoder blocks. The number of blocks is a hyperparameter that can be tuned based on the complexity of the task and the available computational resources. Each block applies a series of transformations to the input sequence to update the representation of each token based on the other tokens in the sequence.

Inside each decoder block, there are two main components:

1. **Masked Multi-Head Self-Attention**: This layer allows each token to attend to all previous tokens in the sequence. The "masked" part refers to the use of a causal mask that prevents each token from attending to future tokens, preserving the autoregressive property. The "multi-head" part refers to performing the self-attention operation multiple times in parallel with different learned parameters, which allows the model to attend to different aspects of the input simultaneously.

2. **Position-wise Feedforward Layer**: This is a simple feedforward neural network that is applied independently to each token. It consists of two linear transformations with a ReLU activation function in between. The

purpose of this layer is to allow the model to transform the representation of each token based on its own features, independent of the other tokens.

In addition to these main components, each decoder block also includes residual connections and layer normalization:

- Residual connections: These are skip connections that add the input of each component to its output. This helps the gradients flow more easily through the network and allows the model to learn identity functions, which can make it easier to stack multiple blocks.

- Layer normalization: This is a technique that normalizes the activations of each layer based on the statistics of the current batch. It helps stabilize the training process and can speed up convergence.

In a language modeling setup, the input sequence and target sequence are typically the same, but the target sequence is shifted one token to the right. For example, if the input sequence is "The cat sat on the," the target sequence would be "cat sat on the <EOS>," where "<EOS>" is a special end-of-sequence token.

During training, the model is fed the input sequence and asked to predict the target sequence. At each step, the model attends to all previous tokens to predict the next token. The causal mask ensures that the model can only attend to previous tokens, not future tokens, which preserves the autoregressive property.

During inference, the model can generate new sequences by iteratively predicting the next token based on the previous tokens. Starting with a prompt sequence, the model generates the next token, appends it to the sequence, and repeats the process until an end-of-sequence token is generated or a maximum length is reached.

The decoder-only transformer has proven to be a powerful architecture for language modeling and other autoregressive tasks. Its ability to capture long-range dependencies and learn complex patterns in the data has led to state-of-the-art results on many benchmarks. However, it also has some limitations, such as the quadratic computational complexity of self-attention with respect to the sequence length, which can make it challenging to scale to very long sequences. Researchers are actively exploring ways to address these limitations, such as using sparse attention mechanisms or hierarchical architectures.

4.2 MASKED MULTI-HEAD SELF-ATTENTION

The self-attention mechanism is the core innovation of the transformer architecture. Let's consider how it works.

The input to the self-attention layer is a sequence of token embeddings. These are dense vector representations of each token that capture its meaning and context. The length of each embedding is a hyperparameter of the model, typically 512 or 1024 in large transformers.

Before the actual self-attention calculation, the input embeddings are linearly projected into three matrices: the query matrix Q, the key matrix K, and the value matrix V. Each of these matrices has the same number of rows as the input sequence, and a number of columns that is a hyperparameter of the model (typically 64 or 128 per attention head). The projection is done using learned weight matrices that are unique to each attention head.

The self-attention operation then computes a weighted sum of the value vectors, where the weights are determined by the compatibility of each value vector's corresponding key vector with the query vector. This can be thought of as each token "querying" all other tokens for relevant information, with the relevance determined by the dot product of the query and key.

Mathematically, this is done by first computing the dot product of the query matrix Q with the transpose of the key matrix K. This gives a matrix of raw attention scores, where each row corresponds to a token in the input sequence, and each column corresponds to a token that it is attending to. The dot product is scaled by the square root of the key dimension to prevent the scores from getting too large, which can lead to vanishing gradients.

Next, a softmax function is applied to each row of the scaled scores. This normalizes the scores so that they sum to 1 for each token, turning them into attention weights that can be interpreted as the relative importance of each other token to the current token.

In the decoder-only transformer, a causal mask is applied to the raw attention scores before the softmax. This is a binary matrix with 1s on and below the diagonal, and 0s above the diagonal. By adding negative infinity to the masked positions (the upper triangle), the softmax output for those positions becomes zero, effectively preventing each token from attending to future tokens. This preserves the autoregressive property of the model.

Finally, the attention weights are multiplied with the value matrix V, and the weighted values are summed for each token. This gives the output of the self-attention layer, which has the same shape as the input embeddings.

Multi-head attention extends this process by performing the self-attention calculation multiple times in parallel, with different learned projection matrices for Q, K, and V in each head. The outputs of all heads are then concatenated and linearly projected to the desired output dimension. This allows the model to attend to different aspects or representations of the input simultaneously, increasing its expressiveness.

After the multi-head attention, the output is passed through a feedforward layer, residual connections, and layer normalization, as described before. This process is repeated for each decoder block in the stack.

The self-attention mechanism is powerful because it allows each token to incorporate information from the entire input sequence, weighted by relevance. This enables the model to capture long-range dependencies and complex relationships between tokens. The multi-head extension further enhances this by allowing the model to attend to different subspaces or representations of the input in parallel.

However, the self-attention operation also has a computational cost that scales quadratically with the sequence length, due to the matrix multiplication between Q and K. This can become prohibitive for very long sequences. Various techniques have been proposed to address this, such as sparse attention, chunking, and approximation methods.

Despite these challenges, the self-attention mechanism has proven to be a powerful and flexible tool for sequence modeling, and has become a cornerstone of modern NLP. Understanding how it works is important to effectively utilizing transformers for various tasks.

4.3 POSITION-WISE FEEDFORWARD LAYER

The position-wise feedforward layer is an important component of the transformer architecture that complements the self-attention mechanism. Let's explore its role and function in more detail.

In each decoder block, after the multi-head self-attention layer, the output is passed through a position-wise feedforward layer. "Position-wise" means that the same feedforward network is applied independently to each position (i.e., each token) in the sequence. In other words, the network processes each token separately and identically, regardless of its position in the sequence.

The feedforward layer typically consists of two linear transformations with a ReLU activation function in between. The first linear transformation expands the input dimension, usually by a factor of 4. This higher-dimensional space allows for more expressive transformations. The ReLU activation then introduces a nonlinearity, enabling the network to learn more complex functions. The second linear transformation then projects the high-dimensional representation back to the original dimension.

Mathematically, for an input vector x at a given position, the feedforward layer computes the following:

$$FFN(x) = max(0, x * W_1 + b_1) * W_2 + b_2$$

W_1 and W_2 are learned weight matrices, b_1 and b_1 are learned bias vectors, and max(0, _) is the ReLU activation function.

The purpose of the feedforward layer is to allow the model to transform the representation of each token based on its own features, independently of the other tokens in the sequence. This is in contrast to the self-attention layer, which computes interactions between all tokens.

The self-attention layer can capture global dependencies, allowing each token to incorporate information from the entire sequence. However, it does not provide a straightforward way to transform the features of each token based on its own content. This is where the position-wise feedforward layer is used.

By applying a nonlinear transformation to each token separately, the feedforward layer can refine and manipulate the token representations based on their individual characteristics. This can help the model capture local patterns and dependencies that may be difficult to learn through self-attention alone.

Furthermore, the expanded dimension in the middle of the feedforward layer allows for a more expressive transformation. The high-dimensional space can embed more complex patterns and relationships, which are then projected back to the original dimension to be compatible with the next layer.

After the feedforward layer, a residual connection and layer normalization are typically applied, just as after the self-attention layer. These help stabilize the network and allow for deeper architectures.

While the position-wise feedforward layer is a simple and effective component, it is not the only possible choice. Some transformer variants experiment with other types of feedforward layers, such as depth-wise convolutions or more complex activation functions. However, the basic two-layer structure with ReLU remains the most common.

In summary, the position-wise feedforward layer in the transformer architecture serves to complement the self-attention mechanism by allowing for local, position-specific transformations of the token features. While the self-attention layer captures global dependencies, the feedforward layer refines the individual token representations based on their own content. Together, these two components enable the transformer to learn rich and expressive sequence representations suitable for a wide range of tasks.

4.4 RESIDUAL CONNECTIONS AND LAYER NORMALIZATION

Residual connections and layer normalization are two important techniques used in the transformer architecture to improve training stability and enable deeper models. Let's consider each of these in more detail.

Residual Connections

Residual connections, also known as *skip connections* or *shortcut connections*, were introduced in the ResNet architecture for computer vision. They have since been widely adopted in various deep learning models, including transformers.

In the transformer, residual connections are used around both the self-attention sublayer and the position-wise feedforward sublayer in each decoder block. This means that the output of each sublayer is added to its input before being passed to the next sublayer.

Mathematically, if the input to a sublayer is x and the function computed by the sublayer is F(x), then the output of the sublayer with a residual connection is as follows:

$$output = F(x) + x$$

The primary motivation for using residual connections is to facilitate the flow of information and gradients through the network. In deep networks, the gradients can sometimes become very small as they are backpropagated through many layers. This is known as the *vanishing gradient problem*, and it can make it difficult to train deep networks effectively.

Residual connections provide a "shortcut" for the gradients to flow directly through the network, bypassing the nonlinear transformations. This helps alleviate the vanishing gradient problem and allows the network to learn identity functions more easily. If a sublayer's optimal function is close to an identity function, it can simply learn a near-zero function, and the residual connection will pass the input through unchanged.

Residual connections allow the model to learn incremental or residual functions with reference to the input. Instead of having to learn a completely new representation at each sublayer, the model can focus on learning the difference or the residual between the input and the desired output. This can be easier to optimize and can lead to faster convergence.

Layer Normalization

Layer normalization is a technique used to normalize the activations of a layer in a deep neural network. It was introduced as an alternative to batch normalization, which is less effective in sequence models like transformers due to the variable sequence lengths and the autoregressive nature of the model.

In the transformer, layer normalization is applied after each residual connection, before passing the output to the next sublayer. Given an input tensor x of shape (batch_size, sequence_length, embedding_dim), layer normalization computes the mean and variance of the embeddings across the embedding dimension for each position in the sequence.

$$mean = 1/embedding\_dim * sum(x, axis = -1)$$
$$variance = 1/embedding\_dim * sum((x - mean)^2, axis = -1)$$

It then normalizes the embeddings by subtracting the mean and dividing by the standard deviation (square root of variance), and applies a learned scaling factor γ and bias β.

$$x_{normalized} = (x - mean) / sqrt(variance + \varepsilon)$$
$$output = \gamma * x_{normalized} + \beta$$

Here, ε is a small constant for numerical stability.

The main purpose of layer normalization is to stabilize the layer outputs and gradients throughout training. By normalizing the activations to have zero mean and unit variance, it helps reduce the internal covariate shift, where the distribution of activations changes during training. This can make the network more robust and easier to train.

Layer normalization also has a regularizing effect, as it limits the range of the activations and prevents them from becoming too large or too small. This can help prevent the exploding or vanishing gradient problems.

Furthermore, by applying normalization separately for each position in the sequence, layer normalization is well-suited to variable-length sequences and autoregressive models like the transformer. It allows the model to maintain a stable activation distribution regardless of the sequence length or the position in the sequence.

Together, residual connections and layer normalization form a powerful combination that enables the transformer to be trained effectively as a deep, multi-layer network. The residual connections help with gradient flow and allow for learning incremental functions, while the layer normalization stabilizes the activations and regularizes the network. These techniques have been crucial to the success of transformers and have inspired similar approaches in other architectures, as well.

There are variations and extensions of these techniques, such as pre-normalization (where the layer normalization is applied before the sub-layer instead of after), or other normalization methods like RMSNorm or PowerNorm. However, the basic principles of residual connections and layer normalization remain fundamental to the transformer architecture and its many derivatives.

4.5 POSITIONAL ENCODING

Positional encoding allows a transformer model to incorporate information about the order and position of tokens in the input sequence. Since the transformer relies solely on attention mechanisms and does not have any inherent notion of order, positional encodings are used to inject this information into the input representations.

The two main types of positional encoding used in transformers are as follows:

1. *Learned positional embeddings* are the most common and widely used form of positional encoding in transformers. In this approach, a separate embedding matrix is learned for positions up to a maximum sequence length, just like the token embeddings.

   For a given maximum sequence length of n, a positional embedding matrix P of shape (n, d_model) is randomly initialized and learned as a parameter of the model. Here, d_model is the dimension of the token embeddings.

   Each position i in the sequence is associated with a unique positional embedding P[i] of dimension d_model. This positional embedding is then added element-wise to the corresponding token embedding E[i] to obtain the final input representation:

   ```
   input[i] = E[i] + P[i]
   ```

   The model learns these positional embeddings during training, allowing it to capture and exploit the positional information in the input sequence. The learned embeddings can encode both absolute positions (i.e., the exact position of each token) and relative positions (i.e., the relative distance between tokens).

   Learned positional embeddings are flexible and have been shown to work well in practice. They allow the model to learn position-dependent patterns and relationships that are specific to the task and dataset at hand.

   One limitation of learned positional embeddings is that they are fixed in size and cannot directly generalize to sequences longer than the maximum length seen during training. If the model needs to process longer sequences at test time, the positional embeddings need to be extrapolated or truncated.

2. *Sinusoidal positional encoding* is an alternative approach that uses sine and cosine functions of different frequencies to create a dense positional code for each position in the sequence.

   In this approach, each dimension of the positional encoding corresponds to a sinusoidal function of a different frequency. For a position i and a dimension j in the positional encoding, the value is computed as follows:

```
PE[i, 2j]   = sin(i / 10000^(2j/d_model))
PE[i, 2j+1] = cos(i / 10000^(2j/d_model))
```

where d_model is the embedding dimension, and 2j and 2j+1 are used to interleave sine and cosine functions.

The main idea of sinusoidal positional encoding is that it creates a unique and deterministic positional code for each position based on a combination of sine and cosine functions with different wavelengths. The wavelengths form a geometric progression from 2π to $10000 \cdot 2\pi$, which allows the model to learn to attend to both short-range and long-range dependencies.

One advantage of sinusoidal positional encoding is that it can extend to sequence lengths longer than seen during training. Since the positional code is defined as a continuous function of position, it can be computed for any arbitrary position, even beyond the maximum length used in training.

Sinusoidal positional encoding is less flexible than learned positional embeddings, as it fixes the relationship between positions and their encodings based on the predefined sinusoidal functions. It may not be able to capture task-specific positional patterns as effectively as learned embeddings.

In practice, learned positional embeddings have become the standard choice in most transformer models, due to their flexibility and strong empirical performance. However, sinusoidal positional encoding remains a viable option, particularly in scenarios where the model needs to handle very long sequences or where the computational cost of learning positional embeddings is a concern.

There are also variations and extensions of positional encoding, such as relative positional encodings, which encode the relative distance between tokens rather than their absolute positions, or hybrid approaches that combine learned and sinusoidal encodings. The choice of positional encoding ultimately depends on the specific requirements and characteristics of the task and dataset.

Regardless of the specific encoding scheme, the main idea is to provide the transformer with explicit information about the order and position of tokens in the input sequence. This enables the self-attention mechanism to incorporate positional information when computing the attention weights and allows the model to capture sequence-dependent patterns and relationships. Positional encoding has been a critical innovation in the transformer architecture and has contributed to its success in a wide range of sequence modeling tasks.

4.6 ROTARY POSITIONAL EMBEDDING (ROPE)

Rotary positional embedding (RoPE) is an advanced technique for incorporating positional information into the transformer architecture. It was introduced as an alternative to learned positional embeddings and sinusoidal positional encoding, aiming to provide a more expressive and efficient way to encode positions.

The main idea behind RoPE is to apply a position-dependent rotation to each embedding vector in the input sequence. Instead of adding a separate positional embedding to the token embeddings, RoPE directly modifies the token embeddings based on their positions.

Here is how RoPE works:

1. Embedding dimension splitting
   The embedding dimension d_model is split into d_model/2 pairs. Each pair consists of two adjacent dimensions in the embedding vector.

2. Pair-wise rotation
   For each pair of dimensions (i, j), a position-dependent rotation matrix R_pos is applied to the corresponding embedding values. The rotation matrix is computed based on the position pos and a pair-specific frequency f_ij:

   ```
   R_pos = [
     [cos(pos * f_ij), -sin(pos * f_ij)],
     [sin(pos * f_ij),  cos(pos * f_ij)]
   ]
   ```

   The frequency f_ij is typically chosen as 1 / 10000^(2(i-1)/d_model), similar to the frequencies used in sinusoidal positional encoding.

3. Rotation application
   The rotation matrix R_pos is applied to each pair of dimensions (i, j) in the embedding vector e_pos at position pos:

   ```
   [e_pos[i], e_pos[j]] = R_pos * [e_pos[i], e_pos[j]]
   ```

   This rotates the embedding values in the pair by an angle determined by the position and the pair-specific frequency.

The rotated embeddings are then used as input to the transformer model, replacing the original token embeddings.

The main advantages of RoPE over other positional encoding methods are as follows:

1. Expressiveness
   Rotations have useful properties like composability and equivariance, which make them more expressive than learned or sinusoidal embeddings. *Composability* means that the rotation for a given position can be composed from the rotations of its constituent parts, enabling the model to capture hierarchical and nested positional relationships. *Equivariance* means that the rotations preserve the relative relationships between embeddings, allowing the model to generalize to unseen positions.

2. Efficiency
   RoPE does not introduce any additional parameters to the model, as the rotations are computed on-the-fly based on the positions and frequencies. This makes it more memory-efficient than learned positional embeddings, which require storing a separate embedding matrix for each position.

3. Long-range sequences
   RoPE can effectively handle very long sequences, as the rotations are defined for any arbitrary position and do not rely on a fixed maximum sequence length. This is particularly useful for tasks involving document-level or context-level sequences that can span thousands of tokens.

4. Spatial and multidimensional data
   RoPE can be extended to higher-dimensional spatial data by applying rotations along each dimension independently. This makes it suitable for tasks like image and video processing, where positional information is crucial.

RoPE has been successfully applied to various natural language processing tasks, including language modeling, machine translation, and text classification. It has become particularly popular in large-scale language models, where the ability to handle long sequences and capture complex positional relationships is critical.

RoPE is a more complex and computationally intensive method compared to learned positional embeddings or sinusoidal encoding. The rotation operations involve trigonometric functions and matrix multiplications, which can be slower to compute than simple addition.

Overall, rotary positional embedding is a powerful and expressive technique for encoding positional information in transformers. It offers a more flexible and efficient alternative to traditional positional encoding methods, particularly for tasks involving long sequences or spatial data. As research in this area continues, we can expect to see further improvements and variations of RoPE in future transformer-based models.

4.7 EFFICIENT ATTENTION WITH ALIBI

The transformer's attention mechanism has quadratic complexity with respect to sequence length, making it expensive for very long sequences. One way to reduce this is by adding an auxiliary attention bias that down-weights distant positions, effectively limiting the attention span.

The Alibi method applies a simple linear bias based on the distance between positions. The bias decreases linearly with distance up to a maximum horizon. This allows the model to attend preferentially to nearby tokens while still having the capacity to draw long-range dependencies when needed.

Alibi can be seen as a "softer" alternative to window-based local attention. It has become a popular addition to the transformer architecture, especially for long sequences.

4.8 TOWARD TRANSFORMERS FOR VISION

While originally proposed for language, the transformer architecture has proven powerful for many other domains, including vision, audio, and multimodal data.

The Vision Transformer (ViT) applies a pure transformer to sequences of image patches for image classification. It first splits an image into patches, linearly embeds each one, adds positional encodings, and then feeds the sequence to a transformer encoder.

The ViT can match or exceed traditional convolutional networks while requiring substantially less inductive bias. This has resulted in increased research into transformer-based models for various vision tasks.

Transformers are also increasingly used for video, which can be seen as a sequence of frames. Video transformers have shown impressive results on tasks like action recognition, video captioning, and video question answering.

4.9 MULTIMODAL TRANSFORMERS

The flexibility of the transformer architecture makes it appealing for multimodal tasks that involve data from multiple modalities, such as vision and language.

A common approach is to use separate transformers to encode each modality, then fuse the encodings with cross-attention to model interactions. This has led to state-of-the-art results on tasks like visual question answering, image captioning, and vision-and-language navigation.

Another direction is to train a single large transformer on multiple modalities jointly. By pretraining on large-scale multimodal data, these models can learn rich cross-modal representations that transfer well to downstream tasks with little or no fine-tuning.

As transformers continue to reveal new capabilities in regards to language and vision tasks, we can expect to see more unified architectures that can handle any kind of data. Language models, vision models, and multimodal models all benefit from this approach. Tools like OpenAI's DALL-E and Google's Imagen, which generate images from text descriptions, show what is possible when different modalities are seamlessly integrated.

By understanding the core components of the transformer, you are now well-equipped to explore transformer-based models across various domains. In the next section, we examine some of the most impactful transformers in NLP and see how they can be applied to real-world language tasks.

4.9.1 Text Mining and Information Extraction

Clinical text mining addresses the challenge of information overload in healthcare, where approximately 80% of patient data exists in unstructured, free-text format (Murdoch & Detsky, 2013). By automatically identifying and extracting relevant concepts, entities, and relationships from clinical text, NLP systems help clinicians rapidly access critical information, enabling more informed decision-making (Wang et al., 2018).

Unstructured clinical text includes a wide range of documents:

- progress notes
- discharge summaries
- radiology reports

- pathology reports
- nursing documentation
- patient-reported symptoms

The ability to extract structured information from these sources creates significant opportunities for clinical decision support, quality improvement, and clinical research. In this section, we explore important clinical text mining tasks and demonstrate practical implementations using Python.

Core Clinical Text Mining Tasks

Named entity recognition (NER) identifies and classifies clinical entities in text, such as diseases, medications, procedures, and anatomical locations (Uzuner et al., 2011). NER serves as the foundation for many downstream tasks and enables extraction of important clinical concepts from narrative text.

```
import spacy
from spacy.language import Language

# Load a pre-trained biomedical NLP model
@Language.factory("clinical_ner")
def create_clinical_ner(nlp, name):
    return nlp.add_pipe("ner")

# Load model trained on biomedical text
try:
    nlp = spacy.load("en_core_sci_md")
except OSError:
    print("Please install the model with: pip install https://s3-us-west-2.amazonaws.com/ai2-s2-scispacy/releases/v0.5.0/en_core_sci_md-0.5.0.tar.gz")
    raise
```

```
# Example clinical text
clinical_note = """Patient presents with a history of
diabetes mellitus and hypertension.
Prescribed metformin 500mg twice daily and lisinopril
10mg once daily.
Patient reports occasional chest pain and shortness of
breath."""

# Process the text
doc = nlp(clinical_note)

# Extract and display entities
print("Named Entities:")
for ent in doc.ents:
    print(f"Text: {ent.text}, Label: {ent.label_}")


# Expected output:
# Named Entities:
# Text: diabetes mellitus, Label: DISEASE
# Text: hypertension, Label: DISEASE
# Text: metformin, Label: CHEMICAL
# Text: 500mg, Label: QUANTITY
# Text: lisinopril, Label: CHEMICAL
# Text: 10mg, Label: QUANTITY
# Text: chest pain, Label: SYMPTOM
# Text: shortness of breath, Label: SYMPTOM
```

Clinical Application: NER enables rapid extraction of critical clinical information from notes, supporting automated problem list generation, medication reconciliation, and clinical coding. For example, a clinical decision support system could automatically flag potential drug-disease contraindications by identifying medications and conditions from the clinical narrative.

Relation Extraction

Relation extraction identifies semantic relationships between named entities, such as drug-disease treatments, adverse events, anatomical location-finding associations, and disease-symptom relationships (Luo et al., 2017). These relationships provide critical context for clinical decision-making.

```
import spacy
import pandas as pd
from spacy.tokens import DocBin
from spacy.tokens import Span

# Load spaCy model
nlp = spacy.load("en_core_web_md")

# Sample clinical text
clinical_note = """
The patient was prescribed metformin for diabetes mellitus.
Lisinopril was provided to manage hypertension and prevent renal complications.
The patient developed a rash after taking amoxicillin, suggesting an allergic reaction.
"""

# Create a custom relation extraction pipeline component
class RelationExtractor:
```

```python
    def __init__(self, nlp):
        # Define relation patterns - simplified for demonstration
        # In production, use a trained model or more sophisticated rules
        self.drug_disease_patterns = [
            {"DRUG": ["prescribed", "given", "taking"], "DISEASE": ["for", "to treat"]},
            {"DRUG": ["to manage", "to control"], "DISEASE": []}
        ]
        self.drug_reaction_patterns = [
            {"DRUG": ["after taking", "following"], "REACTION": ["developed", "experienced"]}
        ]
        # Register entity labels
        if "ner" not in nlp.pipe_names:
            nlp.add_pipe("ner")

    def __call__(self, doc):
        relations = []
        # Simple rule-based approach for demonstration
        # In practice, use a trained model

        # Extract entity spans
        entities = {e.text: (e.start, e.end, e.label_) for e in doc.ents}

        # Identify drugs and diseases
```

```python
        drugs = [e for e in doc.ents if e.label_ in ["CHEMICAL", "DRUG"]]
        diseases = [e for e in doc.ents if e.label_ in ["DISEASE", "PROBLEM"]]
        reactions = [e for e in doc.ents if e.label_ in ["SYMPTOM", "FINDING"]]

        # Simple heuristic: if drug and disease in same sentence, check for treatment relation
        for drug in drugs:
            drug_sent = drug.sent
            for disease in diseases:
                if disease.sent == drug_sent:
                    # Check if "for" or "to treat" appears between drug and disease
                    between_text = doc[drug.end:disease.start].text.lower()
                    if "for" in between_text or "to treat" in between_text:
                        relations.append((drug.text, "TREATS", disease.text))

            # Check for adverse reactions
            for reaction in reactions:
                if reaction.sent == drug_sent:
                    between_text = doc[drug.end:reaction.start].text.lower()
                    if "after" in between_text or "following" in between_text:
                        relations.append((drug.text, "CAUSES", reaction.text))
```

```
        # Add relations to doc._.
        if not doc.has_extension("relations"):
            doc.set_extension("relations", default=[])
        doc._.relations = relations
        return doc

# Add custom labels to spaCy's NER component
ner = nlp.get_pipe("ner")
for label in ["DRUG", "DISEASE", "SYMPTOM", "FINDING",
"REACTION", "PROBLEM"]:
    ner.add_label(label)

# Add our relation extraction component
nlp.add_pipe("relation_extractor",    name="relation_
extractor", last=True)

# Add the relation extractor
@Language.factory("relation_extractor")
def create_relation_component(nlp, name):
    return RelationExtractor(nlp)

# Process text
doc = nlp(clinical_note)

# Add manual entity annotations for demonstration
spans = [
    Span(doc, 5, 6, label="DRUG"),          # metformin
```

```
    Span(doc, 7, 9, label="DISEASE"),        # diabetes mellitus
    Span(doc, 10, 11, label="DRUG"),         # Lisinopril
    Span(doc, 14, 15, label="DISEASE"),      # hypertension
    Span(doc, 17, 19, label="PROBLEM"),      # renal complications
    Span(doc, 24, 25, label="FINDING"),      # rash
    Span(doc, 27, 28, label="DRUG"),         # amoxicillin
    Span(doc, 31, 33, label="REACTION")      # allergic reaction
]
doc.ents = spans


# Extract relations
doc = nlp.get_pipe("relation_extractor")(doc)

# Display extracted relations
print("Extracted Relations:")
for relation in doc._.relations:
    print(f"{relation[0]} {relation[1]} {relation[2]}")

# Manual relation examples (what a fully trained model might extract)
expected_relations = [
    ("metformin", "TREATS", "diabetes mellitus"),
    ("lisinopril", "TREATS", "hypertension"),
```

```
        ("lisinopril", "PREVENTS", "renal complications"),
        ("amoxicillin", "CAUSES", "rash"),
        ("amoxicillin", "CAUSES", "allergic reaction")
]

print("\nExpected Relations (from a fully trained model):")
for relation in expected_relations:
    print(f"{relation[0]} {relation[1]} {relation[2]}")
```

Clinical Application: Relation extraction enables systems to understand complex clinical concepts such as medication indications, adverse drug events, and disease-symptom relationships. A CDSS could alert clinicians to potential adverse drug reactions by identifying historical medication-symptom relationships in patient notes.

Concept Normalization

Concept normalization maps extracted entities to standardized medical terminologies and ontologies such as SNOMED-CT, ICD-10, RxNorm, and LOINC. This process enables semantic interoperability across clinical systems and facilitates integration with structured EHR data.

```
import spacy
from spacy.language import Language

# Define a concept normalization component using UMLS
class UMLSNormalizer:
    def __init__(self, nlp):
```

```python
        # In production, use actual UMLS API or database
        # This is a simplified mock implementation
        self.concept_map = {
            "myocardial infarction": "C0027051",  # UMLS CUI for MI
            "heart attack": "C0027051",           # Same concept, different term
            "mi": "C0027051",                     # Abbreviation
            "atorvastatin": "C0286651",           # UMLS CUI for atorvastatin
            "lipitor": "C0286651",                # Brand name
            "diabetes mellitus": "C0011849",      # UMLS CUI for diabetes
            "diabetes": "C0011849",               # Common form
            "type 2 diabetes": "C0011860",        # More specific type
            "hypertension": "C0020538",           # UMLS CUI for hypertension
            "high blood pressure": "C0020538"     # Alternate term
        }

        # Mapping from UMLS CUI to vocabulary-specific codes
        self.vocabulary_map = {
            "C0027051": {
```

```
                "SNOMED-CT": "22298006",
                "ICD-10": "I21.9",
                "ICD-9": "410.9"
            },
            "C0286651": {
                "RxNorm": "83367",
                "NDC": "0071-0156-23"
            },
            "C0011849": {
                "SNOMED-CT": "73211009",
                "ICD-10": "E11.9",
                "ICD-9": "250.00"
            },
            "C0020538": {
                "SNOMED-CT": "38341003",
                "ICD-10": "I10",
                "ICD-9": "401.9"
            }
        }

    def __call__(self, doc):
        if not doc.has_extension("umls_concepts"):
            doc.set_extension("umls_concepts", default={})

        # Normalize entities to UMLS concepts
        for ent in doc.ents:
```

```python
                entity_text = ent.text.lower()
                if entity_text in self.concept_map:
                    cui = self.concept_map[entity_text]
                    codes = self.vocabulary_map.get(cui, {})
                    doc._.umls_concepts[ent.text] = {
                        "CUI": cui,
                        "VOCABULARIES": codes
                    }

        return doc

# Register the component
@Language.factory("umls_normalizer")
def create_umls_normalizer(nlp, name):
    return UMLSNormalizer(nlp)

# Load spaCy model
nlp = spacy.load("en_core_web_md")

# Add the UMLS normalizer to the pipeline
nlp.add_pipe("umls_normalizer", last=True)

# Example clinical text
clinical_note = """
The patient has been diagnosed with myocardial infarction
and prescribed atorvastatin.
```

```
    Previously, the patient was treated for type 2 diabetes
    and high blood pressure.
    """


    # Process the text
    doc = nlp(clinical_note)


    # Add manual entity annotations for demonstration
    spans = [
        Span(doc, 7, 9, label="PROBLEM"),       #
    myocardial infarction
        Span(doc, 11, 12, label="CHEMICAL"),    #
    atorvastatin
        Span(doc, 19, 21, label="PROBLEM"),     # type 2
    diabetes
        Span(doc, 22, 25, label="PROBLEM")      # high
    blood pressure
    ]
    doc.ents = spans


    # Extract normalized concepts
    print("Normalized Concepts:")
    for ent in doc.ents:
        print(f"Text: {ent.text}")
        if ent.text in doc._.umls_concepts:
            concept = doc._.umls_concepts[ent.text]
            print(f"  UMLS CUI: {concept['CUI']}")
```

```
            print("  Standard Codes:")
            for vocab, code in concept["VOCABULARIES"].
items():
                print(f"    {vocab}: {code}")
        else:
            print("  No normalization available")
```

Clinical Application: Concept normalization enables systems to bridge between various clinical terminologies, facilitating interoperability between systems using different coding standards. This supports functions like problem list reconciliation, cohort identification for clinical trials, and automated billing code assignment.

Temporal Information Extraction

Temporal information extraction identifies and normalizes time-related expressions in clinical text, including event dates, durations, frequencies, and temporal relationships (Sun et al., 2013). This capability is essential for understanding patient timelines and clinical trajectories.

```
import re
import spacy
from spacy.tokens import Span
from spacy.language import Language
from datetime import datetime, timedelta
import dateutil.parser

# Define temporal extraction component
class TemporalExtractor:
```

```python
    def __init__(self, nlp):
        # Regular expressions for different temporal patterns
        self.patterns = {
            "DATE": [
                r"(?:on|dated?|in) (\d{1,2}[-/]\d{1,2}[-/]\d{2,4})",
                r"(?:on|dated?|in) (\d{1,2} (?:Jan|Feb|Mar|Apr|May|Jun|Jul|Aug|Sep|Oct|Nov|Dec)[a-z]* \d{2,4})"
            ],
            "DURATION": [
                r"(?:for|during|over) (\d+) (?:day|week|month|year)s?",
                r"(?:after|following) (\d+) (?:day|week|month|year)s?"
            ],
            "FREQUENCY": [
                r"(\d+ times (?:daily|weekly|monthly|per day|per week|per month))",
                r"(once|twice|three times|four times) (?:daily|weekly|monthly|a day|a week|a month)",
                r"(every (?:day|other day|week|month|hour|morning|evening))",
r"(daily|weekly|monthly|hourly|annually|yearly)"
            ],
            "RELATIVE_TIME": [
                r"(yesterday|today|tomorrow)",
                r"(last|this|next) (?:week|month|year)",
```

```python
                    r"(\d+) (?:day|week|month|year)s? ago"
                ]
            }
            
            # Reference date for computing relative dates
            self.reference_date = datetime.now()
            
            # Initialize normalization dictionaries
            self.normalized_dates = {}
            self.normalized_durations = {}
            self.normalized_frequencies = {}
        
        def __call__(self, doc):
            # Set up entity extensions
            if not doc.has_extension("temporal_entities"):
                doc.set_extension("temporal_entities", default=[])
            
            if not doc.has_extension("temporal_normalized"):
                doc.set_extension("temporal_normalized", default={})
            
            # Extract temporal expressions using regex patterns
            text = doc.text
            spans = []
```

```python
        # Detect temporal expressions
        for label, pattern_list in self.patterns.items():
            for pattern in pattern_list:
                for match in re.finditer(pattern, text, re.IGNORECASE):
                    # Extract the full matched text
                    full_match = match.group(0)
                    # Find character offsets in original text
                    start_char = match.start()
                    end_char = match.end()

                    # Convert character offsets to token positions
                    start_token = None
                    end_token = None
                    for i, token in enumerate(doc):
                        if token.idx <= start_char < token.idx + len(token.text):
                            start_token = i
                        if token.idx <= end_char <= token.idx + len(token.text) and end_token is None:
                            end_token = i + 1

                    if start_token is not None and end_token is not None:
                        # Create a span with the identified label
```

```
                        entity = Span(doc, start_token, end_token, label=label)
                        spans.append(entity)

                        # Attempt to normalize the temporal expression
                        normalized_value = self._normalize_temporal(full_match, label)
                        doc._.temporal_normalized[full_match] = normalized_value

        # Add the extracted temporal entities to the document extensions
        doc._.temporal_entities = spans

        return doc

    def _normalize_temporal(self, temporal_text, label):
        """Normalize temporal expressions to ISO format when possible"""
        try:
            if label == "DATE":
                # Try to convert to ISO date
                try:
                    # Handle "on" or other prefix words
                    date_text = temporal_text.split(" ", 1)[1] if " " in temporal_text else temporal_text
                    parsed_date = dateutil.parser.parse(date_text, fuzzy=True)
```

```python
                    return parsed_date.strftime("%Y-%m-%d")
                except:
                    return temporal_text

        elif label == "DURATION":
            # Extract the numeric value and unit
            match = re.search(r"(\d+) (day|week|month|year)s?", temporal_text, re.IGNORECASE)
            if match:
                value = int(match.group(1))
                unit = match.group(2).lower()
                # Normalize to days
                if unit == "day":
                    days = value
                elif unit == "week":
                    days = value * 7
                elif unit == "month":
                    days = value * 30   # Approximation
                elif unit == "year":
                    days = value * 365  # Approximation
                return f"P{days}D"  # ISO 8601 duration format
            return temporal_text

        elif label == "FREQUENCY":
```

```python
                # Simplified normalization for frequency
            if "daily" in temporal_text or "every day" in temporal_text:
                return "R1/P1D"  # Once per day
            elif "twice daily" in temporal_text or "twice a day" in temporal_text:
                return "R2/P1D"  # Twice per day
            elif "weekly" in temporal_text or "every week" in temporal_text:
                return "R1/P7D"  # Once per week
            elif "monthly" in temporal_text or "every month" in temporal_text:
                return "R1/P1M"  # Once per month
            return temporal_text
        
        elif label == "RELATIVE_TIME":
            # Handle relative temporal expressions
            if "ago" in temporal_text:
                match = re.search(r"(\d+) (day|week|month|year)s? ago", temporal_text)
                if match:
                    value = int(match.group(1))
                    unit = match.group(2).lower()
                    if unit == "day":
                        target_date = self.reference_date - timedelta(days=value)
                    elif unit == "week":
```

```python
                            target_date = self.reference_date - timedelta(weeks=value)
                        elif unit == "month":
                            # Approximation
                            target_date = self.reference_date - timedelta(days=value*30)
                        elif unit == "year":
                            # Approximation
                            target_date = self.reference_date - timedelta(days=value*365)
                        return target_date.strftime("%Y-%m-%d")
                elif "yesterday" in temporal_text:
                    target_date = self.reference_date - timedelta(days=1)
                    return target_date.strftime("%Y-%m-%d")
                elif "today" in temporal_text:
                    return self.reference_date.strftime("%Y-%m-%d")
                elif "tomorrow" in temporal_text:
                    target_date = self.reference_date + timedelta(days=1)
                    return target_date.strftime("%Y-%m-%d")
                return temporal_text
        except:
            # Return the original text if normalization fails
            return temporal_text
```

```python
# Register the component
@Language.factory("temporal_extractor")
def create_temporal_extractor(nlp, name):
    return TemporalExtractor(nlp)


# Load spaCy model
nlp = spacy.load("en_core_web_md")


# Add temporal extraction component to pipeline
nlp.add_pipe("temporal_extractor", last=True)


# Example clinical text
clinical_note = """
The patient experienced chest pain on 01/15/2023 and was admitted to the hospital.
He was discharged after 5 days and prescribed aspirin to be taken twice daily for 6 months.
The patient previously had a similar episode 2 years ago.
Follow-up appointment is scheduled for next week.
"""


# Process the text
doc = nlp(clinical_note)


# Display extracted temporal entities
```

```python
print("Temporal Information:")
for ent in doc._.temporal_entities:
    original_text = ent.text
    normalized = doc._.temporal_normalized.get(original_text, "Not normalized")
    print(f"Text: {original_text}, Label: {ent.label_}, Normalized: {normalized}")


# Example of building a patient timeline
timeline_events = []
for ent in doc._.temporal_entities:
    if ent.label_ == "DATE" and "chest pain" in ent.sent.text:
        timeline_events.append({
            "event": "Chest pain onset",
            "date": doc._.temporal_normalized.get(ent.text, ent.text)
        })
    elif "discharged" in ent.sent.text and ent.label_ == "DURATION":
        # Get admission date
        admission_event = next((e for e in timeline_events if e["event"] == "Chest pain onset"), None)
        if admission_event:
            admission_date = admission_event["date"]
            try:
                # Parse the admission date
```

```python
                    admission_datetime = datetime.strptime(admission_date, "%Y-%m-%d")
                    # Add the duration to get the discharge date
                    duration_match = re.search(r"(\d+) days", ent.text)
                    if duration_match:
                        days = int(duration_match.group(1))
                        discharge_date = (admission_datetime + timedelta(days=days)).strftime("%Y-%m-%d")
                        timeline_events.append({
                            "event": "Hospital discharge",
                            "date": discharge_date
                        })
                except:
                    # Fallback if date parsing fails
                    timeline_events.append({
                        "event": "Hospital discharge",
                        "date": f"{admission_date} + {ent.text}"
                    })

# Display the timeline
print("\nPatient Timeline:")
for event in sorted(timeline_events, key=lambda x: x["date"]):
    print(f"{event['date']}: {event['event']}")
```

Clinical Application: Temporal extraction enables the creation of patient timelines, assessment of treatment adherence, and identification of temporal relationships between events. A CDSS could build a chronological view of a patient's disease progression or treatment response, supporting clinical decision-making by providing context for current symptoms or findings.

Combining Techniques: A Clinical NLP Pipeline

In practice, clinical text mining applications combine multiple techniques to create comprehensive NLP pipelines that extract structured information from clinical narratives. The following code is an example of how the different components can work together.

```
import spacy
from spacy.language import Language
from spacy.tokens import Span
import pandas as pd


# Define a simplified clinical NLP pipeline
class ClinicalNLPPipeline:
    def __init__(self):
        try:
            # Load a pre-trained model
            self.nlp = spacy.load("en_core_web_md")
        except OSError:
            # Fallback to a smaller model
            self.nlp = spacy.load("en_core_web_sm")
            print("Warning: Using a smaller model that may have lower accuracy for clinical text")
```

```python
        # Add custom pipeline components
        self._add_clinical_components()

    def _add_clinical_components(self):
        # This would normally add custom NER models, relation extractors, etc.
        # For demonstration, we'll use rule-based components

        # 1. Add entity ruler for clinical entities
        ruler = self.nlp.add_pipe("entity_ruler", before="ner")
        patterns = [
            {"label": "PROBLEM", "pattern": "diabetes mellitus"},
            {"label": "PROBLEM", "pattern": "hypertension"},
            {"label": "PROBLEM", "pattern": "chest pain"},
            {"label": "MEDICATION", "pattern": "metformin"},
            {"label": "MEDICATION", "pattern": "lisinopril"},
            {"label": "MEDICATION", "pattern": "aspirin"},
            {"label": "MEDICATION", "pattern": "atorvastatin"},
            {"label": "DOSAGE", "pattern": [{"TEXT": {"REGEX": "\\d+"}}, {"TEXT": "mg"}]},
```

```python
            {"label": "FREQUENCY", "pattern": "twice daily"},
            {"label": "FREQUENCY", "pattern": "once daily"},
            {"label": "TEMPORAL", "pattern": [{"TEXT": {"REGEX": "\\d+/\\d+/\\d{4}"}}]},
            {"label": "DURATION", "pattern": [{"TEXT": "for"}, {"TEXT": {"REGEX": "\\d+"}}, {"TEXT": {"IN": ["day", "days", "week", "weeks", "month", "months"]}}]}
        ]
        ruler.add_patterns(patterns)

        # 2. Add a relation extractor (simplified)
        @Language.factory("simple_relation_extractor")
        def create_relation_component(nlp, name):
            return SimpleRelationExtractor(nlp)

        self.nlp.add_pipe("simple_relation_extractor", last=True)

    def process(self, text):
        """Process clinical text and return structured information"""
        doc = self.nlp(text)

        # Extract entities
        entities = [{"text": ent.text, "label": ent.label_, "start": ent.start_char, "end": ent.end_char}
```

```python
                        for ent in doc.ents]
        
        # Extract relations (if available)
        relations = []
        if hasattr(doc._, "relations"):
            relations = doc._.relations
        
        # Extract medication information
        medications = self._extract_medication_info(doc)
        
        # Extract problems/diagnoses
        problems = [ent.text for ent in doc.ents if ent.label_ == "PROBLEM"]
        
        # Return structured data
        return {
            "entities": entities,
            "relations": relations,
            "medications": medications,
            "problems": problems
        }
    
    def _extract_medication_info(self, doc):
        """Extract medication details including dosage, frequency, and duration"""
        medications = []
```

```python
        # Find medication entities
        med_ents = [ent for ent in doc.ents if ent.label_ == "MEDICATION"]

        for med in med_ents:
            med_info = {"name": med.text, "dosage": None, "frequency": None, "duration": None}

            # Find related information in the same sentence
            sent = med.sent

            # Look for dosage
            dosage_ents = [ent for ent in doc.ents if ent.label_ == "DOSAGE" and ent.sent == sent]
            if dosage_ents:
                med_info["dosage"] = dosage_ents[0].text

            # Look for frequency
            freq_ents = [ent for ent in doc.ents if ent.label_ == "FREQUENCY" and ent.sent == sent]
            if freq_ents:
                med_info["frequency"] = freq_ents[0].text

            # Look for duration
            duration_ents = [ent for ent in doc.ents if ent.label_ == "DURATION" and ent.sent == sent]
```

```python
                if duration_ents:
                    med_info["duration"] = duration_ents[0].text
            
            medications.append(med_info)
    
    return medications

# Define a simple relation extractor
class SimpleRelationExtractor:
    def __init__(self, nlp):
        if not Doc.has_extension("relations"):
            Doc.set_extension("relations", default=[])
    
    def __call__(self, doc):
        # Find medication-problem relations
        relations = []
        medications = [ent for ent in doc.ents if ent.label_ == "MEDICATION"]
        problems = [ent for ent in doc.ents if ent.label_ == "PROBLEM"]
        
        for med in medications:
            med_sent = med.sent
            for problem in problems:
                if problem.sent == med_sent:
                    # Check for relation indicators
```

```
                    between_text = doc[med.
end:problem.start].text.lower() if med.end < problem.
start else doc[problem.end:med.start].text.lower()
                    if "for" in between_text or
"to treat" in between_text or "prescribed for" in
between_text:
                        relations.append((med.text,
"TREATS", problem.text))

        doc._.relations = relations
        return doc


# Import Doc class
from spacy.tokens import Doc


# Example usage
pipeline = ClinicalNLPPipeline()


clinical_note = """
Patient presents with a history of diabetes mellitus
and hypertension. Prescribed metformin 500mg twice
daily for diabetes and lisinopril 10mg once daily for
hypertension. Patient reports occasional chest pain on
01/15/2023 and was admitted to the hospital. He was
discharged after 5 days and prescribed aspirin to be
taken twice daily for 6 months.
"""
```

```python
# Process the note
result = pipeline.process(clinical_note)

# Display the structured information
print("Extracted Clinical Information:")
print("\nEntities:")
for entity in result["entities"]:
    print(f"  {entity['text']} ({entity['label']})")

print("\nRelations:")
for relation in result["relations"]:
    print(f"  {relation[0]} {relation[1]} {relation[2]}")

print("\nMedications:")
for med in result["medications"]:
    med_str = f"  {med['name']}"
    if med["dosage"]:
        med_str += f" {med['dosage']}"
    if med["frequency"]:
        med_str += f", {med['frequency']}"
    if med["duration"]:
        med_str
```

The examples in Table 4.1 demonstrate how clinical text mining can help extract structured, actionable information from unstructured clinical text, supporting clinicians in decision-making and improving patient care.

TABLE 4.1 Summary of tasks

Task	Purpose	Example Output
Named Entity Recognition	Identifies and classifies entities in text (e.g., diseases, drugs, and procedures).	"diabetes mellitus" (DISEASE), "metformin" (DRUG)
Relation Extraction	Identifies relationships between entities (e.g., drug-disease relationships and adverse events).	"metformin," "prescribed_for," "diabetes mellitus")
Concept Normalization	Maps extracted entities to medical vocabularies (e.g., SNOMED-CT, ICD-10, and RxNorm).	"myocardial infarction" → UMLS Concept ID: C0027051
Temporal Information	Extracts and normalizes temporal information (e.g., dates, durations, and frequencies).	"on 01/01/2023" (DATE), "after 5 days" (DURATION), "daily for 6 months" (FREQUENCY)

Various NLP techniques have been applied to clinical text mining, ranging from rule-based approaches to machine learning methods (Wang et al., 2018). Rule-based approaches rely on manually crafted linguistic patterns and domain-specific knowledge to extract information, while machine learning methods automatically learn patterns and relationships from annotated training data.

In recent years, deep learning techniques, such as convolutional neural networks (CNNs) and recurrent neural networks (RNNs), have achieved state-of-the-art performance in many clinical text mining tasks (Wu et al., 2020). These models can learn rich, contextual representations of words and sentences, enabling them to capture complex linguistic patterns and manage the ambiguity and variability of clinical language.

There are also significant challenges in applying NLP to clinical text, such as the lack of large-scale, high-quality annotated datasets, the complexity and diversity of clinical language, and the need for domain-specific knowledge and expertise (Demner-Fushman et al., 2009). To address these challenges, researchers are developing novel NLP methods that can learn from limited or noisy data, incorporate medical knowledge and reasoning, and provide interpretable and trustworthy results (Xiao et al., 2018).

Practical Tip: When implementing clinical text mining solutions, carefully consider the specific use case, available data sources, and desired level of accuracy and granularity. Engaging with clinical experts and following established best practices for data preprocessing, annotation, and evaluation can help ensure the quality and reliability of the extracted information (Demner-Fushman et al., 2009). Additionally, using standardized vocabularies and ontologies, such as UMLS (Unified Medical Language System), can facilitate the interoperability and reusability of the text mining results across different clinical systems and applications (Bodenreider, 2004).

4.9.2 Sentiment Analysis and Opinion Mining

Sentiment analysis, also known as *opinion mining*, is another important application of NLP in healthcare. It involves determining the emotional tone, attitude, or opinion expressed in a piece of text, such as positive, negative, or neutral (Liu, 2012). In the clinical domain, sentiment analysis can be used to gauge patient satisfaction, monitor public opinion about health policies, and identify potential mental health issues from social media posts or patient narratives (Denecke & Deng, 2015).

Patient satisfaction is an important indicator of the quality of healthcare services and can have significant impacts on patient outcomes, adherence to treatments, and healthcare costs (Manary et al., 2013). Traditional methods for measuring patient satisfaction, such as surveys and questionnaires, can be time-consuming, costly, and subject to low response rates and recall bias. Sentiment analysis offers a promising alternative by enabling the automatic extraction of patient opinions and emotions from various text sources, such as online reviews, social media posts, and electronic health records (EHRs) (Greaves et al., 2013).

For example, sentiment analysis can be applied to patient reviews on healthcare rating websites, such as Healthgrades and RateMDs, to identify the strengths and weaknesses of healthcare providers and facilities (Gräßer et al., 2018). By analyzing the sentiment expressed in these reviews, healthcare organizations can gain insights into patient experiences, preferences, and expectations, and use this information to improve their services and reputation.

Sentiment analysis can also be used to monitor public opinion and sentiment about health-related topics, such as disease outbreaks, vaccination campaigns, and health policy changes (Sinnenberg et al., 2017). By tracking and analyzing the sentiment expressed in news articles, social media posts, and online discussions, public health agencies and researchers can better understand the public's knowledge, attitudes, and behaviors related to health issues, and develop more effective communication and intervention strategies.

Another potential application of sentiment analysis in healthcare is the detection of mental health issues, such as depression, anxiety, and suicidal ideation, from social media data (Guntuku et al., 2017). By analyzing the sentiment and emotion expressed in individuals' social media posts over time, NLP systems can identify patterns and changes in mood and behavior that may indicate mental health problems, and alert healthcare providers or crisis responders for timely intervention.

Various sentiment analysis techniques have been developed, ranging from rule-based approaches to machine learning methods (Medhat et al., 2014). Rule-based approaches rely on manually crafted sentiment lexicons and linguistic rules to determine the sentiment of a text based on the presence and orientation of sentiment-bearing words and phrases. Machine learning methods, however, automatically learn sentiment patterns and classifications from labeled training data, using algorithms such as support vector machines (SVM), naive Bayes, and deep learning models.

CNNs and RNNs have achieved state-of-the-art performance in sentiment analysis tasks (Zhang et al., 2018). These models can learn rich, contextual representations of words and sentences, capturing complex semantic and syntactic patterns that are predictive of sentiment. Attention mechanisms and memory networks have also been incorporated into deep learning models to improve their interpretability and ability to mange long-range dependencies in text (Tang et al., 2016).

However, there are also challenges in applying sentiment analysis to healthcare, such as the complexity and variability of medical language, the lack of labeled training data, and the need for domain-specific sentiment lexicons and models (Denecke & Deng, 2015). Additionally, sentiment analysis results should be interpreted with caution, as they may not always reflect the true opinions and experiences of patients, and may be biased by factors such as cultural differences, social desirability, and self-selection (Greaves et al., 2013).

Practical Tip: When conducting sentiment analysis on healthcare-related text data, carefully preprocess and normalize the data to address the unique challenges of medical language, such as abbreviations, misspellings, and negations (Denecke & Deng, 2015). Using domain-specific sentiment lexicons and incorporating medical knowledge and context into the sentiment analysis models can also improve their accuracy and relevance (Biyani et al., 2014). Finally, combining sentiment analysis with other NLP techniques, such as topic modeling and named entity recognition, can provide a more comprehensive and nuanced understanding of patient opinions and experiences (Greaves et al., 2013).

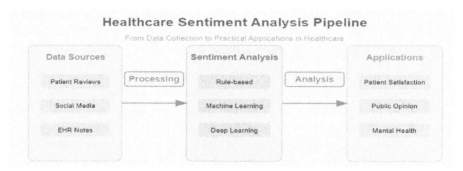

FIGURE 4.1 Healthcare sentiment analysis pipeline

This diagram illustrates the healthcare sentiment analysis pipeline, which transforms raw data into actionable insights. The process begins with Data Sources, including Patient Reviews, Social Media, and EHR Notes. The Processing step applies Sentiment Analysis techniques, which may be Rule-based, Machine Learning, or Deep Learning models. After Analysis, the extracted insights contribute to various Applications, such as Patient Satisfaction, Public Opinion, and Mental Health assessments. This pipeline helps healthcare providers understand patient sentiment and improve healthcare delivery.

4.10 NLP IN CLINICAL DOCUMENTATION

Clinical documentation is critical to healthcare delivery, serving as the primary communication channel among providers, supporting clinical decision-making, and enabling quality improvement and research initiatives (Rosenbloom et al., 2011). However, the exponential growth in documentation requirements has created significant challenges for healthcare providers, who now spend up to 50% of their workday on documentation tasks, often at the expense of direct patient care (Sinsky et al., 2016).

Natural language processing (NLP) offers solutions to streamline clinical documentation workflows and reduce provider burden through automation of key tasks such as information extraction, summarization, and coding (Demner-Fushman et al., 2009). By converting unstructured clinical narratives into structured, actionable data, NLP technologies can simultaneously improve documentation efficiency while enhancing the utility of clinical information for downstream applications.

Important NLP Applications in Clinical Documentation

Medical coding, assigning standardized codes such as ICD-10 (diagnoses) and CPT (procedures), is a critical but resource-intensive process for healthcare organizations. NLP can automate this process by analyzing clinical documentation and suggesting appropriate codes based on documented conditions and procedures (Kaur & Ginige, 2018). This automation can improve coding accuracy, accelerate billing cycles, and reduce the administrative burden on clinicians and coding specialists.

The following example demonstrates a machine learning approach to ICD-10 code prediction from clinical notes.

```
import pandas as pd
import numpy as np
from sklearn.feature_extraction.text import TfidfVectorizer
from sklearn.multioutput import MultiOutputClassifier
from sklearn.linear_model import LogisticRegression
from sklearn.model_selection import train_test_split
from sklearn.metrics import classification_report


# Sample dataset of clinical notes and corresponding ICD-10 codes
# In real applications, this would be a much larger dataset with multiple codes per note
clinical_notes = [
    "Patient with acute bronchitis and fever. Productive cough for 1 week.",
    "Type 2 diabetes poorly controlled with A1C of 9.2. Peripheral neuropathy present.",
    "Chronic obstructive pulmonary disease with acute exacerbation. Increased wheezing and SOB.",
```

```python
    "Hypertension with stage 2 heart failure. Pedal
edema and exercise intolerance.",

    "Acute myocardial infarction, anterior wall.
Severe chest pain radiating to left arm.",

    "Community-acquired pneumonia with hypoxemia
requiring supplemental oxygen.",

    "Urinary tract infection with dysuria and
frequency. Positive leukocyte esterase.",

    "Rheumatoid arthritis with joint swelling and
morning stiffness in hands."
]


# Multiple codes per note (multi-label classification)
icd10_codes = np.array([
    [1, 0, 0, 0, 0],   # J20 (Acute bronchitis)
    [0, 1, 0, 0, 0],   # E11 (Type 2 diabetes)
    [0, 0, 1, 0, 0],   # J44 (COPD)
    [0, 0, 0, 1, 0],   # I10 (Hypertension) + I50
(Heart failure)
    [0, 0, 0, 0, 1],   # I21 (Acute myocardial
infarction)
    [1, 0, 0, 0, 0],   # J18 (Pneumonia)
    [0, 0, 0, 0, 0],   # N39 (UTI)
    [0, 0, 0, 0, 0]    # M06 (Rheumatoid arthritis)
])


# Map indices to actual code descriptions for
readability
```

```python
code_labels = ["Respiratory (J-codes)", "Endocrine (E-codes)", "COPD (J44)",
               "Cardiovascular (I-codes)", "MI (I21)"]


# 1. Create more sophisticated text features using TF-IDF
vectorizer = TfidfVectorizer(
    max_features=1000,
    ngram_range=(1, 2),  # Include both unigrams and bigrams
    stop_words='english',
    min_df=2             # Ignore terms that appear in fewer than 2 documents
)
X = vectorizer.fit_transform(clinical_notes)

# 2. Split into training and testing sets
X_train, X_test, y_train, y_test = train_test_split(
    X, icd10_codes,
    test_size=0.25,
    random_state=42
)

# 3. Train a multi-label classifier
# Using logistic regression as base estimator for each code
base_classifier = LogisticRegression(C=1.0, class_weight='balanced', max_iter=1000)
```

```
multilabel_model = MultiOutputClassifier
(base_classifier)
multilabel_model.fit(X_train, y_train)

# 4. Make predictions
y_pred = multilabel_model.predict(X_test)
y_pred_proba = multilabel_model.predict_proba(X_test)

# 5. Evaluate model performance
print("Model Performance by ICD-10 Code Category:")
for i, code in enumerate(code_labels):
    precision = np.mean(y_test[:, i] == y_pred[:, i])
    print(f"{code}: Accuracy = {precision:.2f}")

# 6. Feature importance analysis
# Identify key terms associated with each code
for i, code in enumerate(code_labels):
    # Get the classifier for this code
    classifier = multilabel_model.estimators_[i]
    # Get feature coefficients
    coefficients = classifier.coef_[0]
    # Get top 5 features with highest coefficients
    top_indices = np.argsort(coefficients)[-5:]
    top_features = [list(vectorizer.vocabulary_.keys())[list(vectorizer.vocabulary_.values()).index(idx)]
```

```
                    for idx in top_indices]
    print(f"\nTop terms for {code}:")
    for feature in reversed(top_features):
        print(f"  - {feature}")


# 7. Demonstrate on a new example
new_note = "Patient presents with shortness of breath, cough, and fever for 3 days. Chest X-ray shows infiltrates consistent with pneumonia."
new_features = vectorizer.transform([new_note])
new_predictions = multilabel_model.predict(new_features)
new_pred_proba = multilabel_model.predict_proba(new_features)


print("\nPredicted ICD-10 Codes for New Patient Note:")
for i, code in enumerate(code_labels):
    if new_predictions[0][i] == 1:
        print(f"  {code} (Confidence: {new_pred_proba[i][0][1]:.2f})")
```

Clinical Application: This automated coding system could be integrated into the EHR to suggest appropriate diagnosis codes as clinicians document patient encounters. The system learns from historical coding patterns and improves over time, while providing confidence scores that help coders prioritize their review efforts. In production settings, such systems typically include post-processing rules to enforce coding guidelines and account for hierarchical relationships between codes.

Clinical Named Entity Recognition

Named entity recognition (NER) automatically identifies and classifies clinical concepts such as diseases, medications, procedures, and lab values from unstructured text. By converting narrative documentation into structured data elements, NER facilitates information retrieval, decision support, and population health management without requiring changes to clinician documentation workflows.

The following example demonstrates a more robust clinical NER implementation.

```
import spacy
from negspacy.negation import Negex
import pandas as pd
from spacy.tokens import Span
from spacy.language import Language


# Custom component to identify medication dosing information
@Language.component("medication_attributes")
def identify_medication_attributes(doc):
    # Extension to store medication attributes
    if not doc.has_extension("medications"):
        doc.set_extension("medications", default=[])

    medications = []

    # Find medication entities
    med_entities = [ent for ent in doc.ents if ent.label_ == "MEDICATION"]
```

```python
    for med_ent in med_entities:
        med_info = {
            "name": med_ent.text,
            "negated": med_ent._.negex,  # Provided by Negex
            "dosage": None,
            "frequency": None,
            "route": None
        }

        # Search for dosage patterns (e.g., "50 mg", "100 mcg")
        dosage_pattern = r'\d+\s*(?:mg|mcg|g|mL)'

        # Search for frequency patterns
        frequency_patterns = [
            "once daily", "twice daily", "three times daily", "four times daily",
            "every morning", "every evening", "every night", "daily", "weekly",
            "every hour", "every 4 hours", "every 6 hours", "every 8 hours",
            "every 12 hours", "as needed", "prn"
        ]

        # Search for route patterns
        route_patterns = [
```

```python
            "by mouth", "orally", "po", "subcutaneous",
"intravenous", "iv",
        "intramuscular", "im", "topical", "sublingual",
"sl", "inhaled"
        ]

        # Check tokens in the same sentence
        sent = med_ent.sent
        sent_text = sent.text.lower()

    # Extract dosage using string matching (simplified approach)
        import re
          dosage_matches = re.findall(dosage_pattern,
sent_text)
        if dosage_matches:
            med_info["dosage"] = dosage_matches[0]

        # Extract frequency
        for pattern in frequency_patterns:
            if pattern in sent_text:
                med_info["frequency"] = pattern
                break

        # Extract route
        for pattern in route_patterns:
            if pattern in sent_text:
```

```python
                med_info["route"] = pattern
                break
        
        medications.append(med_info)
    
    doc._.medications = medications
    return doc


def extract_clinical_entities(text):
    """
    Extract clinical entities from text using a specialized pipeline.
    Returns structured data including problems, medications, and procedures.
    """
    # Load spaCy model
    # In production: use a clinical model like scispaCy
    or a custom-trained model
    nlp = spacy.load("en_core_web_sm")
    
    # Add clinical entity patterns (simplified for demonstration)
    ruler = nlp.add_pipe("entity_ruler", before="ner")
    patterns = [
        {"label": "PROBLEM", "pattern": [{"LOWER": "hypertension"}]},
```

```
            {"label": "PROBLEM", "pattern": [{"LOWER":
"diabetes"}]},
            {"label": "PROBLEM", "pattern": [{"LOWER":
"pneumonia"}]},
            {"label": "PROBLEM", "pattern": [{"LOWER":
"myocardial"}, {"LOWER": "infarction"}]},
            {"label": "PROBLEM", "pattern": [{"LOWER":
"heart"}, {"LOWER": "failure"}]},
        {"label": "MEDICATION", "pattern": [{"LOWER":
"metformin"}]},
            {"label": "MEDICATION", "pattern": [{"LOWER":
"lisinopril"}]},
            {"label": "MEDICATION", "pattern": [{"LOWER":
"aspirin"}]},
            {"label": "MEDICATION", "pattern": [{"LOWER":
"atorvastatin"}]},
          {"label": "PROCEDURE", "pattern": [{"LOWER":
"echocardiogram"}]},
            {"label": "PROCEDURE", "pattern": [{"LOWER":
"x"}, {"LOWER": "ray"}]},
            {"label": "PROCEDURE", "pattern": [{"LOWER":
"mri"}]},
        {"label": "LAB", "pattern": [{"LOWER": "a1c"}]},
        {"label": "LAB", "pattern": [{"LOWER": "ldl"}]},
         {"label": "LAB", "pattern": [{"LOWER": "hdl"}]}
    ]
    ruler.add_patterns(patterns)

    # Add negation detection
```

```
    nlp.add_pipe("negex")

    # Add medication attribute extraction
    nlp.add_pipe("medication_attributes")

    # Process text
    doc = nlp(text)

    # Group entities by type
    entities = {
        "problems": [],
        "medications": doc._.medications,  # From our
custom component
        "procedures": [],
        "labs": []
    }

    # Extract problem entities with negation status
    for ent in doc.ents:
        if ent.label_ == "PROBLEM":
            entities["problems"].append({
                "text": ent.text,
                "negated": ent._.negex
            })
        elif ent.label_ == "PROCEDURE":
            entities["procedures"].append(ent.text)
```

```python
        elif ent.label_ == "LAB":
            entities["labs"].append(ent.text)

    return entities


# Example usage
if __name__ == "__main__":
    clinical_note = """
    Patient is a 65-year-old male with a history of hypertension and type 2 diabetes.

    Current medications include metformin 500 mg twice daily, lisinopril 10 mg once daily, and atorvastatin 20 mg at bedtime.

    The patient denies chest pain but reports occasional shortness of breath.

    Recent labs show A1C of a 7.2 and LDL of 95.

    An echocardiogram was performed last month showing moderate LV dysfunction.

    Patient has no known drug allergies.
    """

    extracted_info = extract_clinical_entities(clinical_note)

    # Display structured information
    print("EXTRACTED CLINICAL INFORMATION\n")
```

```python
    print("Problems:")
    for problem in extracted_info["problems"]:
        status = "NEGATIVE" if problem["negated"] else
"POSITIVE"
        print(f"  - {problem['text']} ({status})")

    print("\nMedications:")
    for med in extracted_info["medications"]:
        med_str = f"  - {med['name']}"
        if med["dosage"]:
            med_str += f", {med['dosage']}"
        if med["frequency"]:
            med_str += f", {med['frequency']}"
        if med["route"]:
            med_str += f", {med['route']}"
        status = "DISCONTINUED" if med["negated"] else
"ACTIVE"
        med_str += f" ({status})"
        print(med_str)

    print("\nProcedures:")
    for procedure in extracted_info["procedures"]:
        print(f"  - {procedure}")

    print("\nLaboratory Tests:")
    for lab in extracted_info["labs"]:
        print(f"  - {lab}")
```

Clinical Application: This NER system could be integrated into an EHR to automatically populate structured data fields from clinical narratives, reducing manual data entry while preserving the natural workflow of clinical documentation. It could also flag missing documentation elements for regulatory compliance or identify potential coding opportunities.

Clinical Note Summarization

As clinical notes grow longer and more complex, summarization technologies help providers quickly understand the most important information about a patient. NLP-based summarization extracts critical details from lengthy documentation, enabling rapid clinical decision-making and facilitating efficient care transitions (Pivovarov & Elhadad, 2015).

The following example demonstrates an extractive summarization approach for clinical notes.

```python
import numpy as np
import networkx as nx
from sklearn.feature_extraction.text import TfidfVectorizer
from sklearn.metrics.pairwise import cosine_similarity
import re


def preprocess_clinical_text(text):
    """Perform basic preprocessing on clinical text"""
    # Split text into sentences
    sentences = re.split(r'(?<!\w\.\w.)(?<![A-Z][a-z]\.)(?<=\.|\?)\s', text)
    sentences = [sentence.strip() for sentence in sentences if sentence.strip()]

    # Clean sentences
```

```python
    clean_sentences = []
    for sentence in sentences:
        # Remove numbers in brackets (often section numbering)
        sentence = re.sub(r'\[\d+\]', '', sentence)
        # Standardize whitespace
        sentence = re.sub(r'\s+', ' ', sentence).strip()
        if sentence:
            clean_sentences.append(sentence)

    return clean_sentences

def text_rank_summarize(text, num_sentences=3):
    """
    Generate a summary using a TextRank-inspired algorithm.
    TextRank is a graph-based ranking algorithm similar to PageRank.
    """
    # Preprocess text
    sentences = preprocess_clinical_text(text)
    if len(sentences) <= num_sentences:
        return sentences

    # Create TF-IDF features
    vectorizer = TfidfVectorizer(stop_words='english')
```

```python
    sentence_vectors = vectorizer.fit_transform(sentences)

    # Create a similarity matrix
  similarity_matrix = cosine_similarity(sentence_vectors)

   # Create a graph and use similarity scores as weights
    nx_graph = nx.from_numpy_array(similarity_matrix)
    scores = nx.pagerank(nx_graph)

    # Rank sentences
     ranked_sentences = sorted(((scores[i], sentence) for i, sentence in enumerate(sentences)),
                              reverse=True)

    # Get top sentences, but preserve original order
       top_indices = sorted([sentences.index(ranked_sentences[i][1]) for i in range(min(num_sentences, len(ranked_sentences)))])
    summary = [sentences[i] for i in top_indices]

    return summary


def clinical_note_summarizer(note, sections=None):
    """
    Summarize a clinical note, with optional section-specific handling.
```

```
    Args:
        note (str): The full clinical note text
        sections (dict, optional): Dictionary of section titles and importance weights
            e.g., {'Assessment': 1.5, 'Plan': 1.5, 'History': 0.8}

    Returns:
        str: Summarized note
    """
    # Default section weights if not provided
    if sections is None:
        sections = {
            'Chief Complaint': 2.0,
            'History of Present Illness': 1.0,
            'Past Medical History': 0.8,
            'Medications': 1.2,
            'Assessment': 1.5,
            'Plan': 1.5,
            'Review of Systems': 0.5
        }

    # Try to identify sections in the note
    section_pattern = r'(?:^|\n)((?:' + '|'.join(map(re.escape, sections.keys())) + r'):)'
    section_matches = list(re.finditer(section_pattern, note, re.IGNORECASE))
```

```python
    # If no sections found, summarize the entire note
    if not section_matches:
        return '\n'.join(text_rank_summarize(note, num_sentences=5))

    # Extract and summarize each section
    summaries = []
    for i, match in enumerate(section_matches):
        section_title = match.group(1)
        section_name = section_title.rstrip(':')
        weight = sections.get(section_name, 1.0)

        # Determine section content
        start_idx = match.end()
        if i < len(section_matches) - 1:
            end_idx = section_matches[i + 1].start()
            section_content = note[start_idx:end_idx].strip()
        else:
            section_content = note[start_idx:].strip()

        # Skip empty sections
        if not section_content:
            continue

        # Determine number of sentences based on weight
        num_sentences = max(1, int(3 * weight))
```

```python
        # Summarize section
        section_summary = text_rank_summarize(section_content, num_sentences=num_sentences)
        if section_summary:
            summaries.append(f"{section_title}")
            summaries.extend([f"- {s}" for s in section_summary])
            summaries.append("")  # Add blank line between sections

    return '\n'.join(summaries)


# Example usage
if __name__ == "__main__":
    sample_note = """
    History of Present Illness:
    The patient is a 62-year-old male with a history of hypertension, type 2 diabetes, and dyslipidemia.
        He presents with shortness of breath that has worsened over the past 2 weeks, especially when
    walking up stairs. He denies chest pain, palpitations, or syncope. Patient has been compliant with
    his medications. He reports mild ankle swelling that improves with elevation. No recent fever or cough.
    He previously had similar symptoms 2 years ago when he was diagnosed with heart failure.


    Past Medical History:
    1. Hypertension, diagnosed 10 years ago
```

2. Type 2 Diabetes, diagnosed 8 years ago

3. Dyslipidemia

4. Heart Failure with reduced ejection fraction (EF 40% in 2021)

5. Gout

Medications:

1. Metformin 1000mg twice daily

2. Lisinopril 20mg daily

3. Atorvastatin 40mg daily

4. Carvedilol 12.5mg twice daily

5. Furosemide 40mg daily, increased to 80mg daily last week by PCP

Physical Exam:

Vitals: BP 142/88, HR 88, RR 18, Temp 98.6F, O2 sat 94% on room air

General: Alert and oriented, no acute distress

HEENT: Normocephalic, atraumatic, moist mucous membranes

Cardiovascular: Irregular rhythm, no murmurs, S3 gallop present

Respiratory: Bilateral crackles at the bases, no wheezing

Abdomen: Soft, non-tender, non-distended

Extremities: 2+ pitting edema bilaterally to mid-calf, no cyanosis

```
    Assessment:
    1. Acute on chronic heart failure exacerbation,
likely precipitated by medication non-adherence and
excessive salt intake
    2. Hypertension, inadequately controlled
    3. Type 2 Diabetes, well-controlled with A1C 6.8%
last month
    4. Stage 3 chronic kidney disease with baseline
creatinine 1.8 mg/dL


    Plan:
    1. Admit to telemetry unit for heart failure
management
    2. Increase furosemide to 80mg IV BID
    3. Continue home medications
    4. Cardiology consult for consideration of device
therapy given EF < 40%
    5. Check BNP, troponin, BMP, CBC in the morning
    6. Begin low sodium diet
    7. Echocardiogram to reassess left ventricular
function
    8. Discharge planning when clinically stable with
heart failure education
    """


    print("ORIGINAL CLINICAL NOTE LENGTH:")
    print(f"{len(sample_note.split())} words\n")
```

```
    note_summary = clinical_note_summarizer(sample_note)

    print("GENERATED SUMMARY:")
    print(note_summary)
    print(f"\nSUMMARY LENGTH: {len(note_summary.split())} words")
    print(f"COMPRESSION RATIO: {len(note_summary.split()) / len(sample_note.split()):.2f}")
```

Clinical Application: This summarization system could be integrated into the EHR to generate concise summaries for handoffs, referrals, or quick chart reviews. The section-aware approach preserves the logical structure of clinical documentation while highlighting the most pertinent information from each section based on clinical priorities.

Clinical Decision Support

NLP can extract relevant information from clinical notes to provide real-time clinical decision support, such as identifying potential drug interactions, suggesting appropriate diagnostic tests or treatments, and alerting providers to critical findings (Demner-Fushman et al., 2009). This can help improve the quality and safety of patient care, while reducing the cognitive burden on healthcare providers.

```
import re
from datetime import datetime

class ClinicalNLPDecisionSupport:
    def __init__(self):
        # Drug interaction database (simplified for demonstration)
```

```python
        self.drug_interactions = {
            frozenset(['warfarin', 'aspirin']): {
                'severity': 'high',
                'description': 'Increased bleeding risk'
            },
            frozenset(['ace inhibitor', 'potassium']): {
                'severity': 'moderate',
                'description': 'Risk of hyperkalemia'
            },
            frozenset(['metformin', 'contrast']): {
                'severity': 'high',
                'description': 'Risk of lactic acidosis; hold metformin 48h after contrast'
            },
            frozenset(['nsaid', 'ssri']): {
                'severity': 'moderate',
                'description': 'Increased bleeding risk'
            },
            frozenset(['statin', 'fibrate']): {
                'severity': 'moderate',
                'description': 'Increased risk of myopathy and rhabdomyolysis'
            }
        }
```

```python
        # Drug class mappings
        self.drug_classes = {
            'lisinopril': 'ace inhibitor',
            'enalapril': 'ace inhibitor',
            'ramipril': 'ace inhibitor',
            'atorvastatin': 'statin',
            'simvastatin': 'statin',
            'rosuvastatin': 'statin',
            'ibuprofen': 'nsaid',
            'naproxen': 'nsaid',
            'celecoxib': 'nsaid',
            'fluoxetine': 'ssri',
            'sertraline': 'ssri',
            'escitalopram': 'ssri',
            'gemfibrozil': 'fibrate',
            'fenofibrate': 'fibrate'
        }

        # Clinical alert rules (simplified)
        self.alert_rules = {
            'medication_missing_dose': r'\b(?:prescribed|started on|continue|taking|start)\s+([A-Za-z]+)\b(?!.*\d+\s*(?:mg|mcg|g|mL|unit))',
            'critical_lab_values': r'\b(?:potassium|K)\s+(?:of\s+)?(?:>|greater than)\s*(?:6\.0|6|7)',
```

```python
            'infection_without_antibiotics': r'\b(?:pneumonia|cellulitis|UTI|urinary tract infection|sepsis)\b',
            'fall_risk': r'\b(?:fall|fell|unsteady|dizzy|syncope)\b',
            'allergy_mention': r'(?:allerg|hypersensitivity|anaphylaxis|allergic)\b'
        }
        
        # Documentation completeness rules
        self.completeness_rules = {
            'allergy_documentation': r'\b(?:allergies|NKDA|no known drug allergies)\b',
            'vital_signs': r'\b(?:vital signs|temperature|blood pressure|BP|pulse|heart rate|respiratory rate|oxygen saturation)\b',
            'physical_exam': r'\b(?:physical exam|examination|HEENT|cardiovascular|respiratory|abdomen|extremities|neurologic)\b',
            'assessment_plan': r'\b(?:assessment|plan|impression|diagnosis)\b'
        }
    
    def detect_drug_interactions(self, text):
        """Identify potential drug interactions in clinical text."""
        interactions = []
        
        # Extract medications from text
        medications = set()
        for drug_name in self.drug_classes.keys():
```

```python
                    if re.search(r'\b' + re.escape(drug_name) + r'\b', text.lower()):
                        medications.add(drug_name)
                        medications.add(self.drug_classes[drug_name])  # Add drug class
        
        # Check for direct medication mentions
        common_drugs = ['warfarin', 'aspirin', 'metformin', 'contrast', 'potassium']
        for drug in common_drugs:
            if re.search(r'\b' + re.escape(drug) + r'\b', text.lower()):
                medications.add(drug)
        
        # Check for interactions
        for interaction_pair, details in self.drug_interactions.items():
            if len(medications.intersection(interaction_pair)) >= 2:
                interactions.append({
                    'medications': list(medications.intersection(interaction_pair)),
                    'severity': details['severity'],
                    'description': details['description']
                })
        
        return interactions
```

```python
    def generate_clinical_alerts(self, text):
        """Generate clinical alerts based on text content."""
        alerts = []
        
        # Check each alert rule
        for alert_type, pattern in self.alert_rules.items():
            matches = re.finditer(pattern, text.lower())
            for match in matches:
                if alert_type == 'medication_missing_dose':
                    medication = match.group(1)
                    alerts.append({
                        'type': 'Medication Missing Dose',
                        'description': f"Medication '{medication}' mentioned without dose specification",
                        'severity': 'low'
                    })
                elif alert_type == 'critical_lab_values':
                    alerts.append({
                        'type': 'Critical Lab Value',
                        'description': f"Possible critical potassium value: {match.group(0)}",
                        'severity': 'high'
                    })
```

```python
                elif alert_type == 'infection_without_antibiotics':
                    infection = match.group(0)
                    # Check if antibiotics are mentioned
                    antibiotics = ['amoxicillin', 'azithromycin', 'ciprofloxacin', 'ceftriaxone', 'vancomycin',
                                   'doxycycline', 'metronidazole', 'penicillin', 'cephalexin']
                    has_antibiotic = any(re.search(r'\b' + re.escape(abx) + r'\b', text.lower()) for abx in antibiotics)
                    if not has_antibiotic:
                        alerts.append({
                            'type': 'Infection Without Antibiotic',
                            'description': f"Possible infection ({infection}) without documented antibiotic therapy",
                            'severity': 'medium'
                        })
                elif alert_type == 'fall_risk':
                    alerts.append({
                        'type': 'Fall Risk',
                        'description': "Patient may be at risk for falls; consider fall precautions",
                        'severity': 'medium'
                    })
```

```python
                elif alert_type == 'allergy_mention':
                    # Check for specific allergens nearby
                    context = text[max(0, match.start() - 50):min(len(text), match.end() + 50)]
                    alerts.append({
                        'type': 'Allergy Alert',
                        'description': f"Allergy mentioned: '{context}'",
                        'severity': 'medium'
                    })
        
        return alerts
    
    def check_documentation_completeness(self, text):
        """Check for completeness of clinical documentation."""
        missing_elements = []
        
        for element, pattern in self.completeness_rules.items():
            if not re.search(pattern, text, re.IGNORECASE):
                missing_elements.append(element.replace('_', ' ').title())
        
        return missing_elements
```

```python
    def analyze_clinical_note(self, text, generate_summary=True):
        """Comprehensive analysis of a clinical note for decision support."""
        results = {
            'drug_interactions': self.detect_drug_interactions(text),
            'clinical_alerts': self.generate_clinical_alerts(text),
            'missing_documentation': self.check_documentation_completeness(text)
        }

        return results


# Example usage
if __name__ == "__main__":
    clinical_note = """
    Patient is a 72-year-old female with history of atrial fibrillation, hypertension, and type 2 diabetes.

    Current medications include warfarin 5mg daily, metformin 1000mg twice daily, and lisinopril 20mg daily.

    Patient reports feeling dizzy when standing up quickly. She fell yesterday but denies head injury.


    Patient is scheduled for CT scan with contrast tomorrow to evaluate abdominal pain.
```

```
    Physical exam shows BP 135/82, HR 78 irregular, RR
16, Temp 98.6F.
    Lungs clear to auscultation. Heart with irregular
rhythm, no murmurs.

    Assessment:
    1. Atrial fibrillation, rate controlled
    2. Orthostatic hypotension
    3. Type 2 diabetes, well-controlled
    4. Abdominal pain, etiology unclear

    Plan:
    1. Continue current medications
    2. Proceed with CT abdomen with contrast tomorrow
    3. Consider reducing lisinopril dose if
orthostatic symptoms persist
    4. Follow-up in 2 weeks
    """

    # Run analysis
    nlp_support = ClinicalNLPDecisionSupport()
    analysis = nlp_support.analyze_clinical_note(clinical_note)

    # Display results
    print("CLINICAL DECISION SUPPORT ALERTS\n")
```

```
    print("Medication Interactions:")
    if analysis['drug_interactions']:
        for interaction in analysis
['drug_interactions']:
            print(f"  ⚠ {interaction['severity'].
upper()} RISK: {', '.join(interaction
['medications'])}")
            print(f"     {interaction
['description']}")
    else:
        print("  No medication interactions detected")
    
    print("\nClinical Alerts:")
    if analysis['clinical_alerts']:
        for alert in analysis['clinical_alerts']:
            severity_symbol = "●" if alert['severity']
== 'high' else "●" if alert['severity'] == 'medium'
else "●"
            print(f"  {severity_symbol}
{alert['type']}: {alert['description']}")
    else:
        print("  No clinical alerts detected")
    
    print("\nDocumentation Gaps:")
    if analysis['missing_documentation']:
        for element in analysis['missing_
documentation']:
            print(f"  ✘ Missing: {element}")
    else:
        print("  Documentation appears complete")
```

Clinical Application: This NLP-powered clinical decision support system could operate in real-time as clinicians document patient encounters, flagging potential safety issues, suggesting relevant interventions, and ensuring documentation completeness. By integrating with the EHR, the system could provide just-in-time guidance without disrupting clinical workflow.

Advanced Techniques for Clinical Documentation NLP

Beyond the basic applications described above, several advanced NLP techniques have emerged that are particularly well-suited to clinical documentation.

Deep Learning Models

Recent advances in deep learning have revolutionized clinical NLP. Transformer-based models like BERT (Bidirectional Encoder Representations from Transformers) and its domain-specific variants such as Clinical BERT and BioBERT have achieved state-of-the-art performance on many clinical NLP tasks (Wu et al., 2020). These models leverage large-scale pre-training on clinical text corpora to develop rich contextual representations of medical language.

The advantages of transformer-based models include the following:

1. Contextual understanding: Understanding that "cold" can have different meanings, such as a "patient has a cold" versus the patient has "cold extremities"
2. Transfer learning: Pretraining on large clinical corpora enables strong performance even with limited task-specific labeled data
3. Multi-task learning: The ability to simultaneously handle multiple NLP tasks such as entity recognition, relation extraction, and text classification

Transformer architectures have demonstrated particularly strong performance in tasks involving complex clinical relationships and temporal reasoning, such as adverse event detection and clinical timeline construction (Jagannatha, 2016).

Hybrid Approaches

While deep learning models offer impressive performance, many practical clinical NLP systems employ hybrid approaches that combine machine learning with rule-based components and domain knowledge. These hybrid systems leverage the strengths of each approach:

1. Rule-based components: Provide precision for well-structured entities (e.g., medication dosing formats, lab values) and enforce clinical constraints
2. Machine learning components: Handle the variability and ambiguity of natural language, discovering patterns that might be difficult to encode as rules
3. Medical knowledge integration: Incorporate ontologies like SNOMED-CT, RxNorm, and UMLS to enhance entity normalization and relationship extraction

Hybrid approaches often achieve better overall performance in real-world clinical settings where interpretability, precision, and domain knowledge integration are essential.

Challenges and Considerations

Despite significant advances, several challenges remain in applying NLP to clinical documentation:

Data Quality and Representation

Clinical documentation varies widely in structure, completeness, and terminology usage across providers and healthcare systems. Common data quality issues include the following:

- Inconsistent terminology: The same clinical concept may be expressed in multiple ways (e.g., "MI" vs. "myocardial infarction" vs. "heart attack").
- Abbreviation ambiguity: Many clinical abbreviations have multiple possible meanings (e.g., "MS" could mean multiple sclerosis, mitral stenosis, or mental status).
- Template-generated text: Modern EHRs often include boilerplate text that may not reflect actual clinical assessment.
- Copy-paste artifacts: Information copied from previous notes may not be updated to reflect current status.

NLP systems must be robust to these variations and ideally should be able to identify potential documentation quality issues.

Privacy and Security

Clinical NLP systems must adhere to strict privacy regulations, such as HIPAA in the United States. Important considerations include the following:

- De-identification: Ensuring that protected health information (PHI) is properly masked or removed before processing
- Secure processing: Implementing appropriate security controls for systems that handle patient data
- Audit trails: Maintaining records of system access and data usage
- Model privacy: Ensuring that NLP models themselves do not memorize or leak sensitive patient information

Clinical Integration and Workflow

For NLP to meaningfully improve clinical documentation, systems must seamlessly integrate into clinical workflows. Important factors include the following:

- Real-time processing: Providing feedback during documentation rather than retrospectively
- EHR integration: Embedding NLP capabilities directly within existing documentation interfaces
- Minimizing cognitive burden: Designing interfaces that present NLP insights without overwhelming clinicians
- Continuous learning: Incorporating clinician feedback to improve system performance over time

Evaluation and Validation

Rigorous evaluation is essential for clinical NLP systems that influence patient care. Evaluation considerations include the following:

- Clinically relevant metrics: Beyond standard NLP metrics like precision and recall, measuring impact on documentation quality, efficiency, and clinical outcomes
- Domain expertise: Involving clinical experts in system evaluation and validation

- Generalizability: Testing performance across different clinical specialties, institutions, and patient populations
- Temporal drift: Monitoring for performance degradation as clinical language and practices evolve

Best Practices for Clinical Documentation NLP

When implementing NLP for clinical documentation, consider the following best practices:

1. Start with well-defined use cases that address specific pain points in documentation workflows, with clearly defined success metrics.
2. Involve clinicians throughout the development process from initial design to testing and deployment.
3. Leverage domain knowledge by incorporating medical ontologies and clinical reasoning into NLP pipelines.
4. Implement robust evaluation frameworks that assess not just technical performance but also clinical utility and impact on workflow.
5. Design for transparency so clinicians understand how NLP-derived insights are generated.
6. Plan for continuous improvement by gathering feedback and periodically retraining models with new data.
7. Consider ethical implications, particularly around bias in training data and potential impacts on clinical decision-making.
8. Start small and scale gradually, beginning with lower-risk applications before advancing to more critical use cases.

By following these practices, healthcare organizations can effectively leverage NLP to enhance clinical documentation, reducing provider burden while simultaneously improving the quality and utility of clinical information.

The Future of NLP in Clinical Documentation

Looking ahead, several emerging trends promise to further expand the role of NLP in clinical documentation:

1. ambient clinical intelligence that can passively listen to provider-patient conversations and generate structured documentation without explicit dictation
2. multimodal learning that combines text analysis with other data modalities such as images, physiologic signals, and structured EHR data
3. personalized documentation models that adapt to individual provider documentation styles and preferences
4. patient-generated text analysis that helps integrate patient-reported outcomes and communications into the clinical record
5. federated learning approaches that enable model training across institutions without sharing sensitive patient data

As these technologies mature, we can expect NLP to move beyond simply processing existing documentation toward actively shaping how clinical information is captured, represented, and utilized across the healthcare ecosystem.

CHAPTER 5

DATA MINING AND PREDICTIVE ANALYTICS

Data mining is the process of discovering patterns, correlations, and anomalies in large datasets to extract valuable insights and knowledge. In healthcare, these techniques are transforming how we understand disease patterns, predict patient outcomes, and allocate resources. *Predictive analytics*, a subfield of data mining, employs statistical and machine learning techniques to forecast future events or outcomes based on historical data.

The healthcare industry generates massive volumes of data daily, from electronic health records and medical imaging to wearable devices and genomic sequencing. By leveraging these rich data sources, predictive analytics can help providers

- make more proactive and personalized clinical decisions
- identify high-risk patients for early intervention
- optimize resource allocation and staffing
- reduce costs while improving care quality
- enhance population health management strategies

This chapter explores techniques and applications of data mining and predictive analytics in healthcare, focusing on association rule mining, clustering, and predictive modeling.

5.1 ASSOCIATION RULE MINING

Association rule mining (ARM) is a powerful technique for discovering interesting relationships between variables in large datasets (Tan et al., 2016; Hahsler et al., 2017). Originally developed for market basket analysis, ARM has found valuable applications in healthcare for identifying disease comorbidities, medication interactions, and treatment patterns.

5.1.1 Theoretical Foundation

Association rule mining identifies relationships in the form of "if-then" rules: if condition A exists, then condition B is likely to occur (Ordonez et al., 2006). These rules are expressed as X → Y, where X (antecedent) and Y (consequent) are disjoint itemsets. The strength of association rules is measured using several metrics:

- Support: The proportion of records containing both X and Y
- Confidence: The conditional probability of Y given X
- Lift: The ratio of observed support to expected support if X and Y were independent
- Conviction: The ratio of expected frequency of X occurring without Y if they were independent, to the observed frequency

For example, a rule {diabetes} → {hypertension} with support of 0.15, confidence of 0.75, and lift of 2.5 indicates that

- 15% of all patients have both diabetes and hypertension.
- 75% of patients with diabetes also have hypertension.
- Patients with diabetes are 2.5 times more likely to have hypertension than the general patient population.

5.1.2 Association Rule Mining Algorithms

Several algorithms have been developed for association rule mining:

Apriori Algorithm: This is the pioneering algorithm that uses breadth-first search and pruning based on the anti-monotone property. If an itemset is infrequent, all its supersets must also be infrequent.

FP-Growth (Frequent Pattern Growth): This creates a compact data structure called an FP-tree and extracts frequent itemsets directly without candidate generation, offering better performance than Apriori for large datasets.

Eclat (Equivalence Class Transformation): This uses a depth-first search approach with set intersection operations, which can be faster than Apriori for certain datasets.

CHARM: This is an efficient algorithm for mining closed frequent itemsets, which can significantly reduce the number of patterns without losing information.

5.1.3 Healthcare Applications of Association Rule Mining

Association rule mining has numerous applications in healthcare:

- Disease Comorbidity Analysis: Identifying conditions that frequently co-occur, such as diabetes and cardiovascular diseases
- Adverse Drug Event Detection: Discovering potential drug-drug interactions and adverse reactions
- Clinical Pathway Analysis: Identifying common treatment patterns and their associations with outcomes
- Fraud Detection: Detecting unusual billing patterns or provider behaviors that may indicate fraudulent practices
- Public Health Surveillance: Analyzing patterns of disease spread and risk factors in populations

5.1.4 Implementation of Association Rule Mining

Let's explore a practical implementation of association rule mining on healthcare data. We use Python with libraries like Pandas, NumPy, and mlxtend to analyze patterns in patient conditions.

```
import pandas as pd
import numpy as np
from mlxtend.frequent_patterns import apriori, association_rules
```

```python
import matplotlib.pyplot as plt
import seaborn as sns

# Configure visualization styles
plt.style.use('seaborn-whitegrid')
sns.set(font_scale=1.2)

# Set random seed for reproducibility
np.random.seed(42)
```

First, we create a synthetic dataset of patient medical conditions that mimics real-world healthcare data.

```python
def generate_synthetic_patient_data(n_patients=2000):
    """
    Generate synthetic patient data with realistic correlations between conditions

    Parameters:
    -----------
    n_patients : int
        Number of patients to generate

    Returns:
    --------
    pd.DataFrame
        DataFrame with binary indicators for each condition
```

```
"""
# Base probabilities for each condition
base_probs = {
    'diabetes': 0.15,            # 15% prevalence
    'hypertension': 0.30,        # 30% prevalence
    'obesity': 0.25,             # 25% prevalence
    'heart_disease': 0.12,       # 12% prevalence
    'high_cholesterol': 0.20,    # 20% prevalence
    'kidney_disease': 0.08,      # 8% prevalence
    'asthma': 0.10,              # 10% prevalence
    'depression': 0.18           # 18% prevalence
}

# Initialize data dictionary with base probabilities
data = {condition: np.random.choice([0, 1], size=n_patients,
                                    p=[1-prob, prob])
        for condition, prob in base_probs.items()}

# Create correlation matrix to represent known medical relationships
for i in range(n_patients):
    # Diabetes increases probability of hypertension, kidney disease, and heart disease
    if data['diabetes'][i] == 1:
```

```python
            if np.random.random() < 0.6:  # 60% of diabetic patients have hypertension
                data['hypertension'][i] = 1
            if np.random.random() < 0.3:  # 30% of diabetic patients have kidney disease
                data['kidney_disease'][i] = 1
            if np.random.random() < 0.4:  # 40% of diabetic patients have heart disease
                data['heart_disease'][i] = 1

        # Obesity increases probability of diabetes, hypertension, and high cholesterol
        if data['obesity'][i] == 1:
            if np.random.random() < 0.35:  # 35% of obese patients have diabetes
                data['diabetes'][i] = 1
            if np.random.random() < 0.55:  # 55% of obese patients have hypertension
                data['hypertension'][i] = 1
            if np.random.random() < 0.45:  # 45% of obese patients have high cholesterol
                data['high_cholesterol'][i] = 1

        # Hypertension increases probability of heart disease
        if data['hypertension'][i] == 1 and np.random.random() < 0.25:
            data['heart_disease'][i] = 1

        # Depression is more common in patients with chronic conditions
```

```python
            chronic_count = sum([data['diabetes'][i],
data['heart_disease'][i],
                                 data['kidney_disease'][i],
data['asthma'][i]])
            if chronic_count >= 2 and np.random.random()
< 0.4:
                data['depression'][i] = 1

    # Convert to DataFrame
    df = pd.DataFrame(data)
    return df


# Generate the dataset
patient_df = generate_synthetic_patient_data(2000)

# Display dataset information
print("Dataset Overview:")
print(f"Number of patients: {len(patient_df)}")
print("\nCondition prevalence (%):")
for column in patient_df.columns:
    prevalence = (patient_df[column].sum() / len(patient_df)) * 100
    print(f"{column}: {prevalence:.1f}%")

# Display sample of the data
print("\nSample of patient data (1=condition present, 0=condition absent):")
print(patient_df.head())
```

Next, we implement the association rule mining process using the Apriori algorithm.

```
def mine_association_rules(df, min_support=0.05, min_confidence=0.3, min_lift=1.5):
    """
    Perform association rule mining on the dataset.

    Parameters:
    -----------
    df : pd.DataFrame
        DataFrame containing binary (0/1) data
    min_support : float
        Minimum support threshold
    min_confidence : float
        Minimum confidence threshold
    min_lift : float
        Minimum lift threshold

    Returns:
    --------
    pd.DataFrame
        DataFrame containing the generated association rules
    """
    # Generate frequent itemsets using the Apriori algorithm
    print(f"Mining frequent itemsets (min_support={min_support})...")
```

```python
    frequent_itemsets = apriori(df,
                                min_support=min_support,
                                use_colnames=True,
                                verbose=1)

    print(f"Found {len(frequent_itemsets)} frequent itemsets")

    # Generate association rules from the frequent itemsets
    print(f"Generating association rules (min_confidence={min_confidence}, min_lift={min_lift})...")
    rules = association_rules(frequent_itemsets,
                              metric="confidence",
                              min_threshold=min_confidence)

    # Filter rules by minimum lift
    rules = rules[rules['lift'] >= min_lift]

    # Sort rules by lift in descending order
    rules = rules.sort_values('lift', ascending=False)

    print(f"Generated {len(rules)} rules meeting the criteria")

    # Calculate additional metrics
    rules['conviction'] = np.where(
```

```
        rules['confidence'] == 1,
        np.inf,
        (1 - rules['consequent support']) / (1 - rules
['confidence'])
    )

    return rules
```

Now, we add functions to analyze and visualize the discovered association rules.

```
def analyze_rules(rules, top_n=10):
    """
    Analyze and display insights from the mined
association rules

    Parameters:
    -----------
    rules : pd.DataFrame
        DataFrame containing association rules
    top_n : int
        Number of top rules to display
    """
    # Handle empty rules case
    if len(rules) == 0:
        print("No rules found meeting the specified
criteria.")
        return
```

```python
    # Display top rules by lift
    print(f"\nTop {min(top_n, len(rules))} rules by lift:")
    pd.set_option('display.max_columns', None)

    # Convert frozensets to strings for better display
    display_rules = rules.copy()
    display_rules['antecedents'] = display_rules['antecedents'].apply(lambda x: ', '.join(list(x)))
    display_rules['consequents'] = display_rules['consequents'].apply(lambda x: ', '.join(list(x)))

    # Display relevant columns with formatted values
    cols = ['antecedents', 'consequents', 'support', 'confidence', 'lift', 'conviction']
    display_cols = display_rules[cols].head(top_n)

    # Format numerical columns
    for col in ['support', 'confidence', 'lift']:
        display_cols[col] = display_cols[col].apply(lambda x: f"{x:.3f}")

    # Handle infinity values in conviction
    display_cols['conviction'] = display_cols['conviction'].apply(
```

```python
        lambda x: "∞" if x == np.inf else f"{x:.3f}"
    )

    print(display_cols)

    # Analyze rule distribution
    print("\nRule Distribution Statistics:")
    numeric_cols = ['support', 'confidence', 'lift']
    for col in numeric_cols:
        print(f"{col.capitalize()}: min={rules[col].min():.3f}, max={rules[col].max():.3f}, mean={rules[col].mean():.3f}")

    # Analyze antecedent items
    all_antecedents = []
    for antecedent in rules['antecedents']:
        all_antecedents.extend(list(antecedent))

    antecedent_counts = pd.Series(all_antecedents).value_counts()
    print("\nTop antecedent items:")
    print(antecedent_counts.head(5))

def visualize_rules(rules, df):
    """
    Create visualizations for the association rules
```

```
Parameters:
-----------
rules : pd.DataFrame
    DataFrame containing association rules
df : pd.DataFrame
    Original dataset (for column names)
"""
if len(rules) == 0:
    print("No rules to visualize.")
    return

# 1. Scatter plot of support vs confidence
plt.figure(figsize=(10, 6))
scatter = plt.scatter(
    rules['support'],
    rules['confidence'],
    alpha=0.8,
    s=rules['lift']*20,  # Size based on lift
    c=rules['lift'],     # Color based on lift
    cmap='viridis'
)

# Add colorbar and labels
plt.colorbar(scatter, label='Lift')
plt.xlabel('Support', fontsize=12)
plt.ylabel('Confidence', fontsize=12)
```

```python
    plt.title('Association Rules: Support vs 
Confidence\n(size and color represent lift)', 
fontsize=14)

    # Add grid for better readability
    plt.grid(True, linestyle='--', alpha=0.7)
    plt.tight_layout()
    plt.show()

    # 2. Network graph of rules (for rules with single 
items)
    try:
        import networkx as nx

        # Filter rules for visualization (single item 
rules are easier to visualize)
        simple_rules = rules[
            (rules['antecedents'].apply(lambda x: 
len(x) == 1)) &
            (rules['consequents'].apply(lambda x: 
len(x) == 1))
        ]

        if len(simple_rules) > 0:
            # Create network graph
            G = nx.DiGraph()

            # Add edges for each rule
```

```
            for _, rule in simple_rules.iterrows():
                antecedent = list(rule['antecedents'])[0]
                consequent = list(rule['consequents'])[0]
                G.add_edge(
                    antecedent,
                    consequent,
                    weight=rule['lift'],
                    confidence=rule['confidence']
                )

            # Set up the plot
            plt.figure(figsize=(12, 10))

            # Position nodes using spring layout
            pos = nx.spring_layout(G, seed=42)

            # Draw the graph
            nx.draw_networkx_nodes(G, pos, node_size=2000, node_color='lightblue', alpha=0.8)

            # Edge weights for thickness
            edge_weights = [G[u][v]['weight'] * 1.5 for u, v in G.edges()]

            # Draw edges with varying thickness based on lift
```

```
                nx.draw_networkx_edges(G, pos, width=edge_
weights, alpha=0.7,
                                                edge_color='gray',
arrows=True, arrowsize=20)


            # Draw labels
            nx.draw_networkx_labels(G, pos, font_
size=12, font_family='sans-serif')


            # Add edge labels (lift values)
            edge_labels = {(u, v): f"L:{G[u][v]
['weight']:.2f}\nC:{G[u][v]['confidence']:.2f}"
                        for u, v in G.edges()}
            nx.draw_networkx_edge_labels(G, pos, edge_
labels=edge_labels, font_size=9)


            plt.title('Network of Association Rules\
n(L: Lift, C: Confidence)', fontsize=14)
            plt.axis('off')
            plt.tight_layout()
            plt.show()
    except ImportError:
        print("NetworkX library not available.
Skipping network visualization.")


    # 3. Heatmap of lift values between conditions
    # Create matrix of lift values for the heatmap
    conditions = list(df.columns)
```

```python
        lift_matrix = pd.DataFrame(0, index=conditions,
columns=conditions)


    # Populate matrix with lift values
    for _, rule in rules.iterrows():
        # Consider only rules with single items for
clarity
        if len(rule['antecedents']) == 1 and
len(rule['consequents']) == 1:
            antecedent = list(rule['antecedents'])[0]
            consequent = list(rule['consequents'])[0]
            lift_matrix.loc[antecedent, consequent] =
rule['lift']


    # Visualize the heatmap if there are values
    if lift_matrix.sum().sum() > 0:
        plt.figure(figsize=(10, 8))
        mask = lift_matrix == 0  # Mask cells with
zero lift (no rule)

        # Create heatmap
        sns.heatmap(
            lift_matrix,
            annot=True,
            cmap='YlOrRd',
            fmt='.2f',
            mask=mask,
```

```
            cbar_kws={'label': 'Lift Value'}
        )

        plt.title('Lift Values Between Conditions', fontsize=14)
        plt.tight_layout()
        plt.show()
```

Finally, we implement the main function to run our association rule mining analysis.

```
def main():
    """Main function to run the association rule mining analysis"""
    # Generate the dataset
    print("Generating synthetic patient data...")
    patient_df = generate_synthetic_patient_data(n_patients=2000)

    # Display dataset information
    print("\nDataset Overview:")
    print(f"Number of patients: {len(patient_df)}")
    print("\nCondition prevalence (%):")
    for column in patient_df.columns:
        prevalence = (patient_df[column].sum() / len(patient_df)) * 100
        print(f"{column}: {prevalence:.1f}%")
```

```python
    # Correlation matrix visualization
    plt.figure(figsize=(10, 8))
    sns.heatmap(patient_df.corr(), annot=True, cmap='coolwarm', fmt='.2f',
                linewidths=0.5, cbar_kws={'label': 'Correlation Coefficient'})
    plt.title('Correlation Matrix of Medical Conditions', fontsize=14)
    plt.tight_layout()
    plt.show()


    # Mine association rules with default parameters
    print("\nMining association rules...")
    rules = mine_association_rules(
        patient_df,
        min_support=0.05,      # At least 5% of patients have this combination
        min_confidence=0.3,    # At least 30% conditional probability
        min_lift=1.5           # At least 50% stronger than random chance
    )


    # Analyze the rules
    analyze_rules(rules, top_n=10)


    # Visualize the rules
    visualize_rules(rules, patient_df)
```

```python
    # Interactive exploration: Allow user to adjust thresholds
    print("\nInteractive Exploration:")
    try:
        support = float(input("Enter minimum support (0.01-0.2, default 0.05): ") or 0.05)
        confidence = float(input("Enter minimum confidence (0.1-0.9, default 0.3): ") or 0.3)
        lift = float(input("Enter minimum lift (1.0-5.0, default 1.5): ") or 1.5)
        
        # Mine rules with user-specified parameters
        custom_rules = mine_association_rules(
            patient_df,
            min_support=support,
            min_confidence=confidence,
            min_lift=lift
        )
        
        # Analyze and visualize custom rules
        analyze_rules(custom_rules, top_n=10)
        visualize_rules(custom_rules, patient_df)
    
    except ValueError:
        print("Invalid input. Using default values.")
    
    print("\nAnalysis complete!")
```

```
# Run the main function
if __name__ == "__main__":
    main()
```

This implementation demonstrates a comprehensive association rule mining pipeline that

1. generates synthetic patient data with realistic correlations between medical conditions
2. applies the Apriori algorithm to identify frequent itemsets
3. generates association rules with support, confidence, and lift metrics
4. analyzes the discovered rules and presents insights
5. visualizes the rules using multiple techniques (scatter plot, network graph, heatmap)
6. allows interactive exploration with adjustable thresholds

5.1.5 Interpreting Association Rules in HealthCare

When interpreting association rules in healthcare, consider the following aspects:

1. Clinical relevance: Not all statistically significant associations are clinically meaningful. Domain experts should evaluate rules for clinical relevance.
2. Directionality vs. causality: Association rules identify relationships but do not establish causality. The rule $X \rightarrow Y$ does not necessarily mean X causes Y.
3. Confounding factors: Relationships might be due to unmeasured variables. For example, an association between two conditions might be due to a common risk factor not captured in the data.
4. Support threshold considerations: Rare but clinically important associations might be missed with high support thresholds. In healthcare, rare events can be critical.
5. Actionability: The most valuable rules are those that suggest actionable interventions or guide clinical decision-making.

5.1.6 Challenges and Limitations

Association rule mining in healthcare faces several challenges:

1. Data quality and completeness: Missing or inaccurate data can lead to misleading associations.
2. High dimensionality: Healthcare datasets often include thousands of variables, making it challenging to identify meaningful patterns.
3. Temporal relationships: Standard association rule mining does not account for the temporal sequence of events, which is crucial in healthcare.
4. Interpretability: Large numbers of rules can be overwhelming for clinicians to interpret and apply.
5. Privacy concerns: Association rule mining on healthcare data must comply with privacy regulations and protect patient confidentiality.

5.1.7 Advanced Techniques for Healthcare Association Rule Mining

Several advanced techniques have been developed to address the limitations of basic association rule mining in healthcare:

1. Temporal Association Rule Mining: Incorporates time relationships between events, such as "medication A followed by symptom B within 7 days"
2. Quantitative Association Rule Mining: Handles numerical variables (like lab values) rather than just binary presence/absence
3. Negative Association Rule Mining: Identifies meaningful absence relationships (e.g., "absence of medication A is associated with complication B")
4. Utility-Based Association Rule Mining: Incorporates the clinical importance or utility of items rather than just their frequency
5. Multilevel Association Rule Mining: Leverages hierarchical medical taxonomies (e.g., ICD-10 codes) to find associations at different levels of granularity

5.1.8 Best Practices for Association Rule Mining in HealthCare

When applying association rule mining to healthcare data, consider these best practices:

1. Domain knowledge integration: Collaborate with clinical experts during all phases, from data preparation to rule interpretation.
2. Careful preprocessing: Handle missing values, outliers, and imbalanced data appropriately.
3. Balanced threshold selection: Choose support, confidence, and lift thresholds that balance statistical significance with clinical relevance.
4. Visualization: Use intuitive visualizations to help clinicians interpret and apply the discovered rules.
5. Validation: Validate discovered associations through statistical tests, literature review, or prospective studies.
6. Iterative refinement: Continuously refine the mining process based on feedback from clinical users.
7. Ethical considerations: Ensure that the application of discovered rules does not perpetuate biases or lead to discriminatory practices (Gianfrancesco et al., 2018).

Association rule mining, when properly applied, can reveal valuable insights from healthcare data that might otherwise remain hidden. These insights can guide clinical decision-making, resource allocation, and population health strategies, ultimately improving patient outcomes and healthcare efficiency.

5.2   CLUSTERING AND SEGMENTATION

Clustering is another important data mining technique that involves grouping similar examples together based on their features or attributes (Berkhin, 2006). In healthcare, clustering can be used to segment patient populations for targeted interventions, identify subtypes of diseases, and discover patterns in patient trajectories or treatment responses (Soni et al., 2011).

Clustering is an unsupervised machine learning technique that identifies natural groupings within data based on similarity. Unlike supervised learning, clustering does not require labeled examples—it discovers inherent structures in the data. In healthcare, clustering enables us to segment patient populations, identify disease subtypes, and discover patterns in treatment responses that might otherwise remain hidden.

The goal of clustering is to partition data into groups where

- examples within the same cluster are more similar to each other
- examples in different clusters are more dissimilar from each other

This similarity is typically measured using distance or similarity functions appropriate for the specific data characteristics.

The goal of clustering is to partition a dataset into groups, or clusters, such that examples within the same cluster are more similar to each other than to examples in other clusters (Xu & Wunsch, 2005). This similarity is typically measured using a distance or similarity function, such as Euclidean distance, cosine similarity, or Jaccard coefficient, depending on the type and scale of the data (Xu & Tian, 2015).

There are various types of clustering algorithms, which can be broadly categorized into the following groups:

1. Partitional clustering: These algorithms divide the dataset into a predetermined number of clusters, typically by optimizing an objective function, such as the sum of squared distances between examples and their cluster centroids. Examples include k-means, k-medoids, and fuzzy c-means clustering.

2. Hierarchical clustering: These algorithms create a hierarchy of clusters, either by recursively merging smaller clusters into larger ones (agglomerative approach) or by recursively splitting larger clusters into smaller ones (divisive approach). Examples include single linkage, complete linkage, and Ward's method.

3. Density-based clustering: These algorithms identify clusters as dense regions of examples separated by regions of lower density. They can discover clusters of arbitrary shape and handle noise and outliers. Examples include DBSCAN (Density-Based Spatial Clustering of Applications with Noise), OPTICS, and Mean Shift.

4. Model-based clustering: These algorithms assume that the data is generated from a mixture of probability distributions and aim to fit the parameters of these distributions to the data. Examples include Gaussian Mixture Models (GMM) and Latent Dirichlet Allocation (LDA).

In healthcare, clustering has been applied to various types of data, such as electronic health records, clinical trial data, and patient-reported outcomes

(Soni et al., 2011). Some examples of healthcare applications of clustering include the following:

1. Patient segmentation: Clustering can be used to group patients with similar clinical characteristics, risk factors, or care needs, enabling personalized interventions and resource allocation (Vuik et al., 2016). For example, clustering patients with diabetes based on their demographics, comorbidities, and medication use can help identify subgroups with different disease trajectories and treatment response.

2. Disease subtyping: Clustering can help identify subtypes of heterogeneous diseases, such as cancer or Alzheimer's disease, based on molecular, imaging, or clinical data (Carreiro et al., 2015). For example, clustering gene expression profiles of breast cancer patients can reveal distinct molecular subtypes with different prognoses and treatment options (Perou et al., 2000).

3. Care pathway analysis: Clustering can be used to discover common patterns or sequences of care events, such as diagnoses, procedures, and medications, from longitudinal healthcare data (Huang et al., 2014). For example, clustering patient trajectories after hospital discharge can identify subgroups with different risks of readmission or adverse outcomes (Hripcsak et al., 2015).

However, there are also challenges and limitations in applying clustering to healthcare data, such as the high dimensionality and sparsity of the data, the presence of noise and missing values, and the lack of ground truth labels for evaluation (Soni et al., 2011). Additionally, the choice of clustering algorithm, distance measure, and number of clusters can have a significant impact on the results, and should be carefully validated using domain knowledge and statistical methods (Xu & Wunsch, 2005).

Practical Tip: When applying clustering to healthcare data, it is important to carefully preprocess and normalize the data to handle the different scales and distributions of the variables (Xu & Tian, 2015). Using dimensionality reduction techniques, such as principal component analysis (PCA) or t-distributed stochastic neighbor embedding (t-SNE), can help visualize and interpret the clustering results in high-dimensional data (Carreiro et al., 2015). Additionally, combining clustering with other data mining techniques, such as association rule mining or classification, can provide a more comprehensive and actionable understanding of the patient subgroups and their characteristics (Soni et al., 2011).

5.2.1 Clinical Applications of Clustering

Clustering techniques have transformed several areas of healthcare:

1. Patient stratification: Identifying subgroups of patients with similar characteristics for targeted interventions and personalized treatment plans
2. Disease subtyping: Discovering distinct variants of diseases based on clinical, genomic, or imaging data, leading to more precise diagnosis and treatment
3. Resource allocation: Grouping patients by resource utilization patterns to optimize staffing, bed management, and service planning
4. Temporal pattern discovery: Uncovering common trajectories in disease progression or recovery to anticipate patient needs
5. Anomaly detection: Identifying unusual cases or outliers that may require special attention or intervention

5.2.2 Types of Clustering Algorithms

Healthcare applications employ various clustering approaches, each with distinct strengths.

Partitional Clustering

Partitional methods divide data into a predetermined number of non-overlapping clusters. The most widely used algorithm is k-means, which minimizes the sum of squared distances between data points and their assigned cluster centers.

Hierarchical Clustering

Hierarchical methods create a tree of clusters (dendrogram) that shows relationships between clusters at different levels. These can be the following:

- Agglomerative (bottom-up): Start with each point as its own cluster, then successively merge the closest clusters.
- Divisive (top-down): Start with all points in one cluster, then recursively split into smaller clusters.

Density-Based Clustering

Density-based algorithms identify clusters as regions of high point density separated by regions of low density. DBSCAN is particularly valuable in healthcare for its ability to

- discover clusters of arbitrary shape
- handle outliers (noise) naturally
- determine the number of clusters automatically

Model-Based Clustering

Model-based approaches assume data is generated from a mixture of probability distributions. Gaussian Mixture Models (GMMs) are commonly used, fitting multiple Gaussian distributions to the data.

5.2.3 Implementing Clustering for Patient Segmentation

Let's implement a practical example of clustering for patient segmentation using Python. We demonstrate multiple clustering approaches on synthetic patient data that represents common healthcare scenarios.

```
import pandas as pd
import numpy as np
import matplotlib.pyplot as plt
import seaborn as sns
from sklearn.preprocessing import StandardScaler
from sklearn.decomposition import PCA
from sklearn.cluster import KMeans, AgglomerativeClustering, DBSCAN
from sklearn.mixture import GaussianMixture
from scipy.cluster.hierarchy import dendrogram, linkage
from sklearn.metrics import silhouette_score
```

```
import warnings
warnings.filterwarnings('ignore')

# Set visualization styles
plt.style.use('seaborn-whitegrid')
sns.set(font_scale=1.2)
```

Generating Synthetic Patient Data

First, we create a synthetic dataset with realistic patient characteristics.

```
def generate_patient_data(n_samples=1000, random_state=42):
    """
    Generate synthetic patient data with multiple features relevant to healthcare

    Parameters:
    -----------
    n_samples : int
        Number of patients to generate
    random_state : int
        Random seed for reproducibility


    Returns:
    --------
    pd.DataFrame
```

```
    Dataframe with patient features
"""
np.random.seed(random_state)


# Create empty dataframe
data = pd.DataFrame()


# Age: multimodal distribution to represent different patient populations
age_young = np.random.normal(loc=30, scale=5, size=int(n_samples*0.3))
age_middle = np.random.normal(loc=50, scale=8, size=int(n_samples*0.4))
age_elderly = np.random.normal(loc=75, scale=7, size=int(n_samples*0.3))
data['age'] = np.concatenate([age_young, age_middle, age_elderly])


# Clip age to realistic range
data['age'] = np.clip(data['age'], 18, 95).astype(int)


# BMI: correlated with age groups
data['bmi'] = np.where(
    data['age'] < 40,
    np.random.normal(loc=24, scale=4, size=n_samples),  # Younger: lower BMI
    np.where(
```

```python
            data['age'] < 65,
            np.random.normal(loc=28, scale=5, size=n_samples),  # Middle-aged: higher BMI
            np.random.normal(loc=26, scale=4, size=n_samples)   # Elderly: moderate BMI
        )
    )
    data['bmi'] = np.clip(data['bmi'], 16, 45)


    # Blood pressure: correlated with age and BMI
    data['systolic_bp'] = 100 + data['age']*0.5 + data['bmi']*0.3 + np.random.normal(0, 10, n_samples)
    data['systolic_bp'] = np.clip(data['systolic_bp'], 90, 200).astype(int)


    # Blood glucose: create distinct diabetic and non-diabetic populations
    # Non-diabetic: normal glucose (70-100 mg/dL)
    # Pre-diabetic: slightly elevated (100-125 mg/dL)
    # Diabetic: high glucose (>125 mg/dL)


    # Base probability of diabetes increases with age and BMI
    diabetes_prob = 0.05 + (data['age'] > 50) * 0.1 + (data['bmi'] > 30) * 0.2
    diabetes_status = np.random.binomial(1, diabetes_prob)
```

```python
    # Assign blood glucose based on diabetes status
    data['blood_glucose'] = np.where(
        diabetes_status == 0,
        # Non-diabetic
        np.random.normal(loc=85, scale=10, size=n_samples),
        # Diabetic
        np.random.normal(loc=140, scale=20, size=n_samples)
    )
    data['blood_glucose'] = np.clip(data['blood_glucose'], 70, 250).astype(int)


    # Cholesterol levels: correlated with age, BMI
    data['total_cholesterol'] = 150 + data['age']*0.6 + data['bmi']*1.0 + np.random.normal(0, 20, n_samples)
    data['total_cholesterol'] = np.clip(data['total_cholesterol'], 120, 300).astype(int)


    # Hospital visits in past year (count): correlated with overall health indicators
    health_risk = (data['age'] > 65).astype(int) + (data['bmi'] > 30).astype(int) + \
                  (data['systolic_bp'] > 140).astype(int) + (data['blood_glucose'] > 125).astype(int)

    data['hospital_visits'] = np.random.poisson(lam=health_risk)
```

```python
    data['hospital_visits'] = np.clip(data['hospital_visits'], 0, 10).astype(int)


    # Medication count: correlated with age and health risk
    data['medication_count'] = np.round(health_risk * 0.8 + data['age']/20 + np.random.normal(0, 0.5, n_samples))
    data['medication_count'] = np.clip(data['medication_count'], 0, 15).astype(int)


    # Depression score (0-27, PHQ-9 scale)
    # Higher score = more severe depression
    # Create correlation with health status and hospital visits
    data['depression_score'] = np.round(health_risk * 0.7 + data['hospital_visits'] * 0.5 + np.random.normal(2, 3, n_samples))
    data['depression_score'] = np.clip(data['depression_score'], 0, 27).astype(int)


    return data


# Generate our dataset
patients = generate_patient_data(n_samples=1000)


# Display summary statistics
print("Patient Dataset Summary:\n")
print(f"Number of patients: {len(patients)}")
```

```python
print("\nFeature Statistics:")
print(patients.describe())

# Visualize distributions of key features
fig, axes = plt.subplots(2, 4, figsize=(20, 10))
axes = axes.flatten()

sns.histplot(patients['age'], kde=True, ax=axes[0])
axes[0].set_title('Age Distribution')

sns.histplot(patients['bmi'], kde=True, ax=axes[1])
axes[1].set_title('BMI Distribution')

sns.histplot(patients['systolic_bp'], kde=True, ax=axes[2])
axes[2].set_title('Systolic BP Distribution')

sns.histplot(patients['blood_glucose'], kde=True, ax=axes[3])
axes[3].set_title('Blood Glucose Distribution')

sns.histplot(patients['total_cholesterol'], kde=True, ax=axes[4])
axes[4].set_title('Total Cholesterol Distribution')

sns.histplot(patients['hospital_visits'], kde=True, ax=axes[5])
```

```
axes[5].set_title('Hospital Visits Distribution')


sns.histplot(patients['medication_count'], kde=True,
ax=axes[6])
axes[6].set_title('Medication Count Distribution')


sns.histplot(patients['depression_score'], kde=True,
ax=axes[7])
axes[7].set_title('Depression Score Distribution')


plt.tight_layout()
plt.show()


# Examine correlations between features
plt.figure(figsize=(10, 8))
correlation_matrix = patients.corr()
sns.heatmap(correlation_matrix, annot=True,
cmap='coolwarm', fmt='.2f', linewidths=0.5)
plt.title('Correlation Matrix of Patient Features')
plt.tight_layout()
plt.show()
```

Data Preprocessing for Clustering

Before applying clustering algorithms, we need to preprocess the data.

```
def preprocess_for_clustering(df):
    """
    Preprocess the patient data for clustering:
    1. Handle missing values
```

```
    2. Standardize features
    3. Apply dimensionality reduction


    Parameters:
    -----------
    df : pd.DataFrame
        Patient dataframe


    Returns:
    --------
    tuple
        (preprocessed_data, scaler, pca_model)
    """
    # Check for missing values
    if df.isnull().sum().sum() > 0:
        print(f"Missing values found: {df.isnull().sum().sum()}")
        df = df.fillna(df.mean())

    # Standardize the features
    scaler = StandardScaler()
    scaled_data = scaler.fit_transform(df)

    # Apply PCA for dimensionality reduction and visualization
    pca = PCA(n_components=2)
    pca_result = pca.fit_transform(scaled_data)
```

```
    print(f"Explained variance ratio: {pca.
explained_variance_ratio_}")
    print(f"Total explained variance: {sum(pca.
explained_variance_ratio_):.2f}")

    return scaled_data, pca_result, scaler, pca


# Preprocess the data
scaled_data, pca_result, scaler, pca =
preprocess_for_clustering(patients)


# Visualize the PCA projection
plt.figure(figsize=(10, 8))
plt.scatter(pca_result[:, 0], pca_result[:, 1],
alpha=0.6)
plt.title('PCA Projection of Patient Data')
plt.xlabel('Principal Component 1')
plt.ylabel('Principal Component 2')
plt.grid(True)
plt.show()
```

K-Means Clustering

We start with k-means, a simple but effective partitional clustering algorithm.

```
def apply_kmeans(data, pca_result, max_clusters=10):
    """
    Apply K-means clustering with optimal K selection
using the elbow method
```

    and silhouette analysis

    Parameters:
    ----
    data : array-like
        Preprocessed data for clustering
    pca_result : array-like
        PCA projection for visualization
    max_clusters : int
        Maximum number of clusters to evaluate

    Returns:
    ----
    tuple
        (optimal_k, cluster_labels)
    """
    # Calculate inertia (sum of squared distances to nearest centroid)
    inertias = []
    silhouette_scores = []

    for k in range(2, max_clusters + 1):
        kmeans = KMeans(n_clusters=k, random_state=42, n_init=10)
        kmeans.fit(data)
        inertias.append(kmeans.inertia_)

```python
        # Calculate silhouette score
        labels = kmeans.labels_
        silhouette_scores.append(silhouette_score(data, labels))

    # Plot elbow method
    plt.figure(figsize=(18, 6))

    plt.subplot(1, 2, 1)
    plt.plot(range(2, max_clusters + 1), inertias, marker='o')
    plt.title('Elbow Method for Optimal K')
    plt.xlabel('Number of Clusters (K)')
    plt.ylabel('Inertia')
    plt.grid(True)

    # Plot silhouette scores
    plt.subplot(1, 2, 2)
    plt.plot(range(2, max_clusters + 1), silhouette_scores, marker='o')
    plt.title('Silhouette Score for Optimal K')
    plt.xlabel('Number of Clusters (K)')
    plt.ylabel('Silhouette Score')
    plt.grid(True)

    plt.tight_layout()
    plt.show()
```

```python
    # Determine optimal K
    # For simplicity, we'll choose K with highest silhouette score
    optimal_k = silhouette_scores.index(max(silhouette_scores)) + 2
    print(f"Optimal number of clusters based on silhouette score: {optimal_k}")


    # Apply K-means with optimal K
    final_kmeans = KMeans(n_clusters=optimal_k, random_state=42, n_init=10)
    cluster_labels = final_kmeans.fit_predict(data)


    # Visualize final clustering result using PCA projection
    plt.figure(figsize=(12, 10))
    plt.scatter(pca_result[:, 0], pca_result[:, 1], c=cluster_labels,
                cmap='viridis', s=50, alpha=0.8)


    # Plot centroids
    centroids_pca = pca.transform(scaler.inverse_transform(final_kmeans.cluster_centers_))
    plt.scatter(centroids_pca[:, 0], centroids_pca[:, 1], c='red', marker='X', s=200, alpha=1)


    plt.title(f'K-means Clustering (K={optimal_k})')
    plt.xlabel('Principal Component 1')
    plt.ylabel('Principal Component 2')
```

```
    plt.colorbar(label='Cluster')
    plt.grid(True)
    plt.show()

    return optimal_k, cluster_labels, final_kmeans.
cluster_centers_

# Apply K-means clustering
k_optimal, kmeans_labels, kmeans_centers = apply_
kmeans(scaled_data, pca_result, max_clusters=10)
```

Analyzing K-means Clusters

After identifying clusters, we need to characterize them to derive clinical insights.

```
def analyze_clusters(df, labels, centers=None, 
scaler=None, algorithm_name=""):
    """
    Analyze the characteristics of each cluster

    Parameters:
    -----------
    df : pd.DataFrame
        Original patient dataframe
    labels : array-like
        Cluster assignments
    centers : array-like, optional
        Cluster centers (for K-means)
```

```
    scaler : object, optional
        Scaler used for preprocessing
    algorithm_name : str
        Name of clustering algorithm
    """
    # Add cluster labels to original dataframe
    df_with_clusters = df.copy()
    df_with_clusters['cluster'] = labels

    # Get number of clusters
    n_clusters = len(set(labels))
    if -1 in labels:  # For DBSCAN which assigns -1 to noise points
        n_clusters -= 1

    # Get cluster sizes
    cluster_sizes = df_with_clusters['cluster'].value_counts().sort_index()

    # Plot cluster sizes
    plt.figure(figsize=(12, 6))
    cluster_sizes.plot(kind='bar')
    plt.title(f'Cluster Sizes - {algorithm_name}')
    plt.xlabel('Cluster')
    plt.ylabel('Number of Patients')
    plt.grid(True, axis='y')
    plt.show()
```

```python
    # Calculate mean values for each feature by cluster
    cluster_means = df_with_clusters.groupby('cluster').mean()

    # Plot heatmap of feature means by cluster
    plt.figure(figsize=(14, 8))
    sns.heatmap(cluster_means, annot=True, cmap='YlGnBu', fmt='.1f')
    plt.title(f'Cluster Characteristics - {algorithm_name}')
    plt.tight_layout()
    plt.show()

    # Compare cluster centers with original features (for K-means)
    if centers is not None and scaler is not None:
        # Transform centers back to original scale
        original_centers = scaler.inverse_transform(centers)
        centers_df = pd.DataFrame(original_centers, columns=df.columns)
        centers_df.index.name = 'cluster'

        # Plot radar chart for each cluster
        # Set up the figure
        num_clusters = len(centers_df)
        num_features = len(df.columns)
```

```python
    # Create angles for radar chart
    angles = np.linspace(0, 2*np.pi, num_features, endpoint=False).tolist()
    angles += angles[:1]  # Close the circle

    # Normalize the centers data for radar chart
    normalized_centers = centers_df.copy()
    for col in centers_df.columns:
        normalized_centers[col] = (centers_df[col] - df[col].min()) / (df[col].max() - df[col].min())

    fig, ax = plt.subplots(figsize=(15, 10), subplot_kw=dict(polar=True))
    feature_names = df.columns.tolist()
    feature_names += feature_names[:1]  # Close the circle

    # Plot each cluster
    for i, row in normalized_centers.iterrows():
        values = row.values.tolist()
        values += values[:1]  # Close the circle
        ax.plot(angles, values, linewidth=2, label=f'Cluster {i}')
        ax.fill(angles, values, alpha=0.1)

    # Set feature labels
    ax.set_xticks(angles[:-1])
    ax.set_xticklabels(feature_names[:-1])
```

```python
        # Add legend and title
        plt.legend(loc='upper right', bbox_to_
anchor=(0.1, 0.1))
        plt.title(f'Cluster Profiles - {algorithm_
name}', size=15)
        plt.show()


    # Detailed statistics for each cluster
    for cluster in range(-1, n_clusters):
        if cluster == -1 and -1 not in labels:
            continue


        cluster_data = df_with_clusters[df_with_
clusters['cluster'] == cluster]
        print(f"\n{'=' * 50}")
        print(f"Cluster {cluster}: {len(cluster_data)} 
patients ({len(cluster_data)/len(df)*100:.1f}% of 
total)")
        print(f"{'-' * 50}")


        # Calculate mean, std, min, max for each 
feature
        stats = cluster_data.drop('cluster', axis=1).
describe().loc[['mean', 'std', 'min', 'max']].T
        stats['mean_diff'] = stats['mean'] - df.mean()
        stats['importance'] = abs(stats['mean_diff'] / 
df.std())
        stats = stats.sort_values('importance', 
ascending=False)
```

```python
        print("Key differentiating features (sorted by importance):")
        for feature, row in stats.iterrows():
            direction = "higher" if row['mean_diff'] > 0 else "lower"
            print(f"- {feature}: {row['mean']:.1f} ({direction} than population mean of {df[feature].mean():.1f} by {abs(row['mean_diff']):.1f})")

        # Clinical interpretation
        interpretations = []

        # Age-based interpretation
        if cluster_data['age'].mean() > df['age'].mean() + 10:
            interpretations.append("Elderly patient group")
        elif cluster_data['age'].mean() < df['age'].mean() - 10:
            interpretations.append("Younger patient group")

        # Medical condition interpretations
        if cluster_data['systolic_bp'].mean() > 140:
            interpretations.append("Hypertensive")
        if cluster_data['blood_glucose'].mean() > 125:
            interpretations.append("Likely diabetic")
        if cluster_data['bmi'].mean() > 30:
            interpretations.append("Obese")
```

```python
            if cluster_data['depression_score'].mean() > 10:
                interpretations.append("Moderate-to-severe depression")
            
            # Healthcare utilization
            if cluster_data['hospital_visits'].mean() > 3:
                interpretations.append("High healthcare utilizers")
            if cluster_data['medication_count'].mean() > 5:
                interpretations.append("Multiple medications (polypharmacy)")
            
            print("\nPossible clinical profile:")
            if interpretations:
                for i, interp in enumerate(interpretations):
                    print(f"{i+1}. {interp}")
            else:
                print("Average health status patients without distinctive characteristics")
    
    return df_with_clusters


# Analyze K-means clusters
patients_with_kmeans = analyze_clusters(patients, kmeans_labels,
```

```
                                centers=kmeans_centers,
                                                           scaler=scaler,
algorithm_name="K-means")
```

Hierarchical Clustering

Next, we implement hierarchical clustering to identify nested relationships.

```
def apply_hierarchical_clustering(data, pca_result, n_clusters=None):
    """
    Apply hierarchical clustering and visualize the dendrogram

    Parameters:
    -----------
    data : array-like
        Preprocessed data for clustering
    pca_result : array-like
        PCA projection for visualization
    n_clusters : int, optional
        Number of clusters to cut the dendrogram

    Returns:
    --------
    array-like
        Cluster labels
    """
```

```python
    # Compute linkage matrix for dendrogram
    print("Computing linkage matrix (this may take a while for large datasets)...")
    linkage_matrix = linkage(data, method='ward')

    # Plot dendrogram
    plt.figure(figsize=(14, 8))
    dendrogram(linkage_matrix, truncate_mode='level', p=5)
    plt.title('Hierarchical Clustering Dendrogram')
    plt.xlabel('Sample index')
    plt.ylabel('Distance')
    plt.axhline(y=20, c='k', ls='--', lw=1)
    plt.show()

    # If n_clusters is not provided, use the same as K-means for comparison
    if n_clusters is None:
        n_clusters = k_optimal

    # Apply hierarchical clustering with the specified number of clusters
    hierarchical = AgglomerativeClustering(n_clusters=n_clusters, linkage='ward')
    hierarch_labels = hierarchical.fit_predict(data)

    # Visualize hierarchical clustering result
    plt.figure(figsize=(12, 10))
```

```
    scatter = plt.scatter(pca_result[:, 0], pca_
result[:, 1], c=hierarch_labels,
                cmap='viridis', s=50, alpha=0.8)
    plt.title(f'Hierarchical Clustering 
(n_clusters={n_clusters})')
    plt.xlabel('Principal Component 1')
    plt.ylabel('Principal Component 2')
    plt.colorbar(scatter, label='Cluster')
    plt.grid(True)
    plt.show()
    
    return hierarch_labels

# Apply hierarchical clustering
hierarch_labels = apply_hierarchical_
clustering(scaled_data, pca_result, 
n_clusters=k_optimal)

# Analyze hierarchical clusters
patients_with_hierarch = analyze_clusters(patients, 
hierarch_labels,
algorithm_name="Hierarchical Clustering")
```

DBSCAN: Density-Based Clustering

DBSCAN is particularly valuable for identifying irregularly shaped clusters and outliers.

```
def apply_dbscan(data, pca_result):
    """
    Apply DBSCAN clustering with parameter tuning
```

```
    Parameters:
    -----------
    data : array-like
        Preprocessed data for clustering
    pca_result : array-like
        PCA projection for visualization

    Returns:
    --------
    array-like
        Cluster labels
    """
    # Find optimal eps parameter by calculating k-distance graph
    from sklearn.neighbors import NearestNeighbors

    neigh = NearestNeighbors(n_neighbors=5)
    neigh.fit(data)
    distances, indices = neigh.kneighbors(data)

    # Sort distances for k-distance graph
    distances = np.sort(distances[:, 4])

    # Plot k-distance graph
    plt.figure(figsize=(10, 6))
    plt.plot(distances)
    plt.title('K-distance Graph (k=5)')
```

```python
    plt.xlabel('Points sorted by distance')
    plt.ylabel('5th Nearest Neighbor Distance')
    plt.grid(True)

    # Find the "elbow" point - approximate by taking a point where the curve increases rapidly
    # For demonstration, we'll use a simple percentile approach
    elbow_index = int(len(distances) * 0.95)  # Use 95th percentile as a heuristic
    eps_value = distances[elbow_index]
    plt.axhline(y=eps_value, color='r', linestyle='--')
    plt.text(len(distances)*0.5, eps_value*1.1, f'Suggested eps = {eps_value:.2f}')
    plt.show()

    # Apply DBSCAN with the determined eps value
    print(f"Applying DBSCAN with eps={eps_value:.2f}, min_samples=5")
    dbscan = DBSCAN(eps=eps_value, min_samples=5)
    dbscan_labels = dbscan.fit_predict(data)

    # Count number of clusters and noise points
    n_clusters = len(set(dbscan_labels)) - (1 if -1 in dbscan_labels else 0)
    n_noise = list(dbscan_labels).count(-1)
    print(f"DBSCAN identified {n_clusters} clusters and {n_noise} noise points")
```

```python
    # Visualize DBSCAN clustering result
    plt.figure(figsize=(12, 10))
    scatter = plt.scatter(pca_result[:, 0], pca_result[:, 1], c=dbscan_labels,
                cmap='viridis', s=50, alpha=0.8)
    plt.title(f'DBSCAN Clustering (eps={eps_value:.2f}, min_samples=5)\n{n_clusters} clusters, {n_noise} noise points')
    plt.xlabel('Principal Component 1')
    plt.ylabel('Principal Component 2')
    plt.colorbar(scatter, label='Cluster')
    plt.grid(True)
    plt.show()

    return dbscan_labels


# Apply DBSCAN clustering
dbscan_labels = apply_dbscan(scaled_data, pca_result)

# Analyze DBSCAN clusters
patients_with_dbscan = analyze_clusters(patients, dbscan_labels,

algorithm_name="DBSCAN")
```

Gaussian Mixture Models (GMM)

GMMs provide a probabilistic approach to clustering, modeling data as coming from a mixture of Gaussian distributions.

```python
def apply_gmm(data, pca_result, max_components=10):
    """
    Apply Gaussian Mixture Model with BIC for optimal component selection

    Parameters:
    -----------
    data : array-like
        Preprocessed data for clustering
    pca_result : array-like
        PCA projection for visualization
    max_components : int
        Maximum number of components to evaluate

    Returns:
    --------
    tuple
        (optimal_n_components, cluster_labels)
    """
    # Calculate BIC for different numbers of components
    n_components_range = range(1, max_components + 1)
    bic_scores = []

    for n_components in n_components_range:
        gmm = GaussianMixture(n_components=n_components,
```

```python
                            covariance_type='full',
                            random_state=42,
                            n_init=10)
    gmm.fit(data)
    bic_scores.append(gmm.bic(data))

# Plot BIC scores
plt.figure(figsize=(10, 6))
plt.plot(n_components_range, bic_scores, marker='o')
plt.title('BIC Score by Number of Components')
plt.xlabel('Number of Components')
plt.ylabel('BIC Score')
plt.grid(True)
plt.show()

# Find optimal number of components (lowest BIC)
optimal_n_components = n_components_range[np.argmin(bic_scores)]
print(f"Optimal number of components based on BIC: {optimal_n_components}")

# Apply GMM with optimal number of components
final_gmm = GaussianMixture(n_components=optimal_n_components,
                            covariance_type='full',
                            random_state=42,
```

```python
                            n_init=10)
    final_gmm.fit(data)
    gmm_labels = final_gmm.predict(data)


    # Visualize GMM clustering result
    plt.figure(figsize=(12, 10))
    scatter = plt.scatter(pca_result[:, 0], pca_result[:, 1], c=gmm_labels,
              cmap='viridis', s=50, alpha=0.8)


    # Plot contours of the GMM components if 2D PCA explains sufficient variance
    if sum(pca.explained_variance_ratio_) > 0.5:
        from matplotlib.patches import Ellipse

        # Display the Gaussian mixtures
        w_factor = 0.2 / final_gmm.weights_.max()
        for pos, covar, w in zip(final_gmm.means_, final_gmm.covariances_, final_gmm.weights_):
            # Project the Gaussian to PCA space
            pos_pca = pca.transform([scaler.inverse_transform([pos])])[0]

            # Get the eigenvalues and eigenvectors of covariance
            U, s, Vt = np.linalg.svd(covar)
            angle = np.degrees(np.arctan2(U[0, 1], U[0, 0]))
```

```python
            width, height = 2 * np.sqrt(s)

            # Draw the ellipse (simplified for visualization)
            ellipse = Ellipse(
                xy=pos_pca,
                width=width * w_factor * 5,
                height=height * w_factor * 5,
                angle=angle,
                alpha=0.3
            )
            plt.gca().add_patch(ellipse)

    plt.title(f'Gaussian Mixture Model (n_components={optimal_n_components})')
    plt.xlabel('Principal Component 1')
    plt.ylabel('Principal Component 2')
    plt.colorbar(scatter, label='Cluster')
    plt.grid(True)
    plt.show()

    # Calculate cluster membership probabilities
    membership_probs = final_gmm.predict_proba(data)

    # Identify uncertain patients (similar probabilities for multiple clusters)
    max_probs = np.max(membership_probs, axis=1)
```

```python
    uncertain_threshold = 0.6  # Define patients with less than 60% max probability as uncertain
    uncertain_patients = np.where(max_probs < uncertain_threshold)[0]

    print(f"Identified {len(uncertain_patients)} patients with uncertain cluster assignments")

    # Visualize cluster probabilities for a sample of patients
    sample_size = min(10, len(uncertain_patients))
    if sample_size > 0:
        sample_indices = np.random.choice(uncertain_patients, sample_size, replace=False)

        plt.figure(figsize=(12, sample_size * 0.8))
        for i, idx in enumerate(sample_indices):
            plt.subplot(sample_size, 1, i+1)
            plt.bar(range(optimal_n_components), membership_probs[idx])
            plt.title(f'Patient {idx} Cluster Probabilities')
            plt.xlabel('Cluster')
            plt.ylabel('Probability')
        plt.tight_layout()
        plt.show()

    return optimal_n_components, gmm_labels
```

```
# Apply GMM clustering
gmm_components, gmm_labels = apply_gmm(scaled_data, 
pca_result, max_components=10)


# Analyze GMM clusters
patients_with_gmm = analyze_clusters(patients, 
gmm_labels, 
                                      algorithm_
name="Gaussian Mixture Model")
```

Comparing Clustering Results

After applying different clustering algorithms, we compare their results.

```
def compare_clustering_results(df, clustering_results, 
pca_result):
    """
    Compare different clustering algorithms
    
    Parameters:
    ----------
    df : pd.DataFrame
        Original patient dataframe
    clustering_results : dict
        Dictionary of clustering results with 
algorithm names as keys
    pca_result : array-like
        PCA projection for visualization
    """
```

```python
    # Calculate adjusted Rand index for each pair of clustering results
    from sklearn.metrics import adjusted_rand_score

    algorithms = list(clustering_results.keys())
    n_algorithms = len(algorithms)

    # Create matrix for adjusted Rand scores
    rand_matrix = np.zeros((n_algorithms, n_algorithms))

    for i, alg1 in enumerate(algorithms):
        for j, alg2 in enumerate(algorithms):
            if i == j:
                rand_matrix[i, j] = 1.0
            else:
                # Handle potential noise points in DBSCAN (-1 labels)
                labels1 = clustering_results[alg1]
                labels2 = clustering_results[alg2]

                # If both algorithms have valid labels for all points
                if -1 not in labels1 and -1 not in labels2:
                    rand_matrix[i, j] = adjusted_rand_score(labels1, labels2)
                else:
```

```python
                        # Filter out noise points (keep only points that have valid clusters in both)
                        valid_indices = np.where((labels1 != -1) & (labels2 != -1))[0]
                        if len(valid_indices) > 0:
                            rand_matrix[i, j] = adjusted_rand_score(
                                labels1[valid_indices],
                                labels2[valid_indices]
                            )
                        else:
                            rand_matrix[i, j] = np.nan

    # Visualize the adjusted Rand score matrix
    plt.figure(figsize=(10, 8))
    sns.heatmap(rand_matrix, annot=True, cmap='YlGnBu', fmt='.3f',
                xticklabels=algorithms, yticklabels=algorithms)
    plt.title('Adjusted Rand Index Between Clustering Algorithms')
    plt.tight_layout()
    plt.show()

    # Visualize all clustering results side by side
    fig, axes = plt.subplots(2, 2, figsize=(20, 16))
    axes = axes.flatten()
```

```python
    for i, (algorithm, labels) in 
enumerate(clustering_results.items()):
        scatter = axes[i].scatter(pca_result[:, 0], 
pca_result[:, 1],
                                   c=labels, 
cmap='viridis', s=50, alpha=0.8)
        axes[i].set_title(f'Clustering: {algorithm}')
        axes[i].set_xlabel('Principal Component 1')
        axes[i].set_ylabel('Principal Component 2')
        axes[i].grid(True)
        plt.colorbar(scatter, ax=axes[i], 
label='Cluster')

    plt.tight_layout()
    plt.show()


    # Count patients assigned to the same cluster 
across algorithms
    # Create a unique ID for each cluster assignment 
pattern
    df_comparison = df.copy()


    for algorithm, labels in clustering_results.
items():
        df_comparison[f'cluster_{algorithm}'] = labels


    # Create a pattern string for each patient's 
cluster assignments
```

```python
    df_comparison['cluster_pattern'] = df_comparison.apply(
        lambda row: '_'.join([str(row[f'cluster_{alg}']) for alg in algorithms]),
        axis=1
    )

    # Count pattern frequencies
    pattern_counts = df_comparison['cluster_pattern'].value_counts().head(10)

    print("Top cluster assignment patterns across algorithms:")
    for pattern, count in pattern_counts.items():
        pattern_clusters = pattern.split('_')
        print(f"Pattern: {' | '.join([f'{alg}:{cls}' for alg, cls in zip(algorithms, pattern_clusters)])}")
        print(f"Count: {count} patients ({count/len(df)*100:.1f}%)")
        print('-' * 50)

    return df_comparison

# Create dictionary of clustering results
clustering_results = {
    'K-means': kmeans_labels,
    'Hierarchical': hierarch_labels,
    'DBSCAN': dbscan_labels,
```

```
    'GMM': gmm_labels
}


# Compare clustering results
comparison_df = compare_clustering_results(patients,
clustering_results, pca_result)
```

Clinical Interpretation and Application

The final step is deriving actionable insights from our clustering analysis.

```
def clinical_interpretation(df_with_clusters,
algorithm="K-means"):
    """
    Provide clinical interpretation of clustering results with
    interventions and care strategies for each patient group

    Parameters:
    -----------
    df_with_clusters : pd.DataFrame
        Patient dataframe with cluster assignments
    algorithm : str
        Clustering algorithm to use for interpretations
    """
    cluster_column = f'cluster_{algorithm}'
    if cluster_column not in df_with_clusters.columns:
```

```python
        cluster_column = 'cluster'  # Fallback

    # Number of clusters
    clusters = sorted(df_with_clusters[cluster_
column].unique())
    if -1 in clusters:  # Remove noise points from
DBSCAN
        clusters.remove(-1)

    print(f"\n{'='*80}")
    print(f"CLINICAL INTERPRETATION OF {algorithm.
upper()} CLUSTERING RESULTS")
    print(f"{'='*80}")

    for cluster in clusters:
        cluster_patients = df_with_clusters[df_with_
clusters[cluster_column] == cluster]

        print(f"\n{'-'*80}")
        print(f"Cluster {cluster}: {len(cluster_
patients)} patients ({len(cluster_patients)/
len(df_with_clusters)*100:.1f}%)")
        print(f"{'-'*80}")

        # Calculate mean values for key features
        age_mean = cluster_patients['age'].mean()
        bmi_mean = cluster_patients['bmi'].mean()
```

```python
        sbp_mean = cluster_patients['systolic_bp'].mean()
        glucose_mean = cluster_patients['blood_glucose'].mean()
        cholesterol_mean = cluster_patients['total_cholesterol'].mean()
        hospital_mean = cluster_patients['hospital_visits'].mean()
        meds_mean = cluster_patients['medication_count'].mean()
        depression_mean = cluster_patients['depression_score'].mean()

        # Determine clinical profile
        print("Clinical Profile:")

        # Age group
        if age_mean < 40:
            print("- Young adult patient group")
        elif age_mean < 65:
            print("- Middle-aged patient group")
        else:
            print("- Elderly patient group")

        # BMI category
        if bmi_mean < 18.5:
            print("- Underweight (BMI < 18.5)")
        elif bmi_mean < 25:
```

```python
        print("- Normal weight (BMI 18.5-24.9)")
    elif bmi_mean < 30:
        print("- Overweight (BMI 25-29.9)")
    else:
        print("- Obese (BMI ≥ 30)")

    # Blood pressure
    if sbp_mean < 120:
        print("- Normal blood pressure")
    elif sbp_mean < 130:
        print("- Elevated blood pressure")
    elif sbp_mean < 140:
        print("- Stage 1 hypertension")
    else:
        print("- Stage 2 hypertension")

    # Blood glucose
    if glucose_mean < 100:
        print("- Normal blood glucose")
    elif glucose_mean < 126:
        print("- Prediabetes")
    else:
        print("- Diabetes")

    # Cholesterol
    if cholesterol_mean < 200:
```

```python
            print("- Desirable cholesterol levels")
        elif cholesterol_mean < 240:
            print("- Borderline high cholesterol")
        else:
            print("- High cholesterol")

        # Healthcare utilization
        if hospital_mean < 1:
            print("- Low healthcare utilization")
        elif hospital_mean < 3:
            print("- Moderate healthcare utilization")
        else:
            print("- High healthcare utilization")

        # Medication burden
        if meds_mean < 3:
            print("- Low medication burden")
        elif meds_mean < 6:
            print("- Moderate medication burden")
        else:
            print("- High medication burden (polypharmacy)")

        # Depression severity
        if depression_mean < 5:
            print("- Minimal or no depression")
```

```python
        elif depression_mean < 10:
            print("- Mild depression")
        elif depression_mean < 15:
            print("- Moderate depression")
        elif depression_mean < 20:
            print("- Moderately severe depression")
        else:
            print("- Severe depression")

        # Determine key risks
        print("\nKey Risk Factors:")
        risks = []

        if sbp_mean >= 140:
            risks.append("Uncontrolled hypertension")
        if glucose_mean >= 126:
            risks.append("Uncontrolled diabetes")
        if bmi_mean >= 30:
            risks.append("Obesity")
        if cholesterol_mean >= 240:
            risks.append("Hypercholesterolemia")
        if depression_mean >= 15:
            risks.append("Significant depression")
        if meds_mean >= 6:
            risks.append("Polypharmacy (risk of drug interactions)")
        if age_mean >= 75 and hospital_mean >= 2:
```

```python
                risks.append("Elderly with frequent hospitalizations")
        
        if risks:
            for risk in risks:
                print(f"- {risk}")
        else:
            print("- No major risk factors identified")
        
        # Suggested interventions
        print("\nRecommended Interventions:")
        
        # Age-specific interventions
        if age_mean >= 65:
            print("- Geriatric assessment for fall risk and functional status")
            print("- Medication review to minimize inappropriate prescribing")
        
        # BMI interventions
        if bmi_mean >= 30:
            print("- Weight management program")
            print("- Nutritional counseling")
            print("- Consider referral to bariatric medicine")
```

```python
        # Blood pressure interventions
        if sbp_mean >= 130:
            print("- Blood pressure monitoring and management")
            if sbp_mean >= 140:
                print("- Intensification of antihypertensive therapy")

        # Glucose interventions
        if glucose_mean >= 126:
            print("- Diabetes management and education")
            print("- Regular HbA1c monitoring")
            if glucose_mean > 180:
                print("- Consider endocrinology referral for intensive management")

        # Cholesterol interventions
        if cholesterol_mean >= 200:
            print("- Lipid management")
            if cholesterol_mean >= 240:
                print("- Consider statin therapy or intensification")

        # Depression interventions
        if depression_mean >= 10:
            print("- Depression screening and monitoring")
```

```python
        if depression_mean >= 15:
            print("- Mental health referral")
            print("- Consider pharmacotherapy and/or psychotherapy")
        
        # Healthcare utilization interventions
        if hospital_mean >= 3:
            print("- Care coordination services")
            print("- Transitional care management")
            print("- Remote patient monitoring")
        
        # Medication burden interventions
        if meds_mean >= 5:
            print("- Comprehensive medication review")
            print("- Medication reconciliation")
            if meds_mean >= 8:
                print("- Deprescribing initiatives")
        
        # Care model suggestions
        print("\nSuggested Care Model:")
        
        # High-complexity patients
        if (hospital_mean >= 2 and meds_mean >= 5) or (len(risks) >= 3):
            print("- High-intensity care management")
            print("- Multidisciplinary care team")
```

```python
            print("- Frequent monitoring (biweekly or monthly)")
            print("- Consider home-based primary care if mobility issues")
        
        # Moderate-complexity patients
        elif (hospital_mean >= 1 or meds_mean >= 3) or (len(risks) >= 1):
            print("- Moderate-intensity care management")
            print("- Quarterly follow-up appointments")
            print("- Disease-specific education programs")
            print("- Consider telehealth monitoring")
        
        # Low-complexity patients
        else:
            print("- Standard primary care")
            print("- Annual wellness visits")
            print("- Preventive care focus")
            print("- Self-management education")
        
        print("\n")

# Apply clinical interpretation to K-means clustering results
clinical_interpretation(comparison_df, algorithm="K-means")
```

5.2.4 Clinical Applications of Clustering Results

The clusters identified through these algorithms provide valuable insights for several healthcare applications:

1. Personalized Care Planning

    Patient clusters can be used to develop targeted care plans:

    - High-risk elderly cluster: These patients may benefit from comprehensive geriatric assessments, medication reviews, and care coordination services.
    - Young diabetic cluster: Early intervention programs, lifestyle modification coaching, and frequent monitoring could prevent disease progression.
    - Depression with chronic disease cluster: Integrated behavioral health and chronic disease management programs can address both physical and mental health needs.

2. Resource Allocation

    Healthcare organizations can optimize resource allocation based on clustering results:

    - High utilization clusters: Deploy care managers, social workers, and community health workers to reduce preventable healthcare utilization.
    - Polypharmacy clusters: Allocate clinical pharmacists for medication therapy management services.
    - Low-risk clusters: Implement less resource-intensive monitoring through telehealth and patient self-management tools.

3. Predictive Risk Stratification

    Cluster assignments can enhance predictive models for various healthcare outcomes:

    - Hospital readmission risk: Certain clusters may have significantly higher readmission rates, warranting proactive interventions.
    - Medication adherence: Demographic and clinical profiles of clusters may predict adherence challenges.
    - Disease progression: Clusters may exhibit different trajectories in disease course, informing preventive strategies.

4. Population Health Management

    At a system level, clustering provides insights for population health initiatives:

    - Geographic distribution: Mapping cluster prevalence across regions can identify areas needing targeted resources.
    - Preventive service gaps: Identify clusters with low preventive service utilization for focused outreach campaigns.
    - Social determinants: Correlate cluster membership with social determinants of health to design holistic interventions.

5.2.5 Challenges and Considerations in Healthcare Clustering

While clustering offers powerful insights, several challenges must be addressed.

Data Quality and Preparation

- Missing values: Healthcare data often contains missing values that can significantly impact clustering results. Multiple imputation or model-based approaches may be needed.
- Feature selection: Including irrelevant features can obscure meaningful patterns. Domain knowledge should guide feature selection and engineering.
- Standardization: Features must be appropriately scaled to prevent variables with larger ranges from dominating the analysis.

Algorithm Selection

- No single best algorithm: Different clustering algorithms capture different aspects of the data structure. Multiple approaches should be compared.
- Parameter tuning: Results are sensitive to parameter choices (e.g., number of clusters, distance metrics). Systematic tuning is essential.
- Validation: Without ground truth labels, validating clusters requires clinical interpretation and stability analysis.

Interpretability and Implementation

- Clinical relevance: Statistically-sound clusters may not always be clinically meaningful. Domain experts must validate and interpret results.

- Actionability: Clusters should suggest concrete interventions or care strategies to be useful in practice.
- Integration with workflows: Clustering insights must be integrated into clinical workflows to impact care delivery.

5.2.6 Best Practices for Healthcare Clustering

To maximize the value of clustering in healthcare, consider these best practices:

1. Start with clear objectives: Define specific questions that clustering should address (e.g., "Identify distinct patient subgroups for tailored interventions").
2. Incorporate domain knowledge: Involve clinicians in feature selection, algorithm choice, and result interpretation.
3. Use multiple algorithms: Apply several clustering approaches and compare results for robustness.
4. Validate clusters clinically: Verify that clusters represent meaningful patient subgroups with distinct clinical needs.
5. Visualize results: Use dimensionality reduction and visualization techniques to communicate findings effectively.
6. Combine with predictive models: Use cluster assignments as features in predictive models for outcomes of interest.
7. Implement longitudinal analysis: Track cluster stability and transitions over time to understand patient trajectories.
8. Assess intervention effectiveness: Measure how cluster-specific interventions impact outcomes and refine approaches.

5.2.7 Future Directions

Emerging approaches in healthcare clustering include the following:

1. Dynamic clustering: Capturing how patients move between clusters over time rather than assigning static clusters
2. Multi-view clustering: Integrating diverse data modalities (clinical, genomic, social) for more comprehensive patient subtyping

3. Deep clustering: Using deep learning methods to identify complex patterns in high-dimensional healthcare data
4. Explainable clustering: Developing methods that provide transparent explanations for cluster assignments

Clustering will continue to play a crucial role in advancing precision medicine and population health by enabling more personalized and efficient healthcare delivery tailored to the unique needs of distinct patient subgroups.

5.3 PREDICTIVE MODELING IN PATIENT CARE

Predictive analytics involves using data mining and machine learning techniques to make predictions about future events or outcomes based on historical data (Steyerberg, 2019). In healthcare, predictive modeling can be used to identify patients at risk of developing certain conditions, predict treatment responses and adverse events, and optimize resource allocation and care delivery (Parikh et al., 2019).

The goal of predictive modeling is to learn a function that maps input features (e.g., patient characteristics, clinical variables) to output variables (e.g., disease onset, hospital readmission) based on a labeled training dataset. This function can then be used to make predictions on new, unseen data points (Kuhn & Johnson, 2019).

There are various types of predictive models used in healthcare, depending on the nature of the input and output variables and the specific task at hand (Goldstein et al., 2017):

1. Classification models: These models predict a categorical output variable, such as the presence or absence of a disease, or the type of treatment response. Examples include logistic regression, decision trees, random forests, and support vector machines.
2. Regression models: These models predict a continuous output variable, such as the expected length of hospital stay or the probability of survival. Examples include linear regression, polynomial regression, and Cox proportional hazards models.
3. Time-to-event models: These models predict the time until an event occurs, such as the onset of a disease or the readmission to a hospital, while accounting for censoring and competing risks. Examples include survival analysis methods, such as Kaplan-Meier estimators and Cox regression.

4. Multilevel models: These models account for the hierarchical or clustered structure of healthcare data, such as patients nested within providers or hospitals. They can estimate the effects of individual and group-level factors on patient outcomes. Examples include mixed-effects models and hierarchical Bayesian models.

Predictive modeling in healthcare typically involves the following steps (Kuhn & Johnson, 2019):

1. Data preprocessing: This step involves cleaning, integrating, and transforming the raw healthcare data into a suitable format for modeling. It may include handling missing values, outliers, and inconsistencies, as well as feature selection and engineering.

2. Model training: This step involves fitting the predictive model to the training data using an appropriate algorithm and hyperparameter settings. It may also involve techniques such as cross-validation and regularization to prevent overfitting and improve generalization.

3. Model evaluation: This step involves assessing the performance of the trained model on a separate test dataset using relevant metrics, such as accuracy, precision, recall, and area under the receiver operating characteristic curve (AUC-ROC). It may also involve comparing different models and selecting the best one based on their performance and interpretability.

4. Model deployment: This step involves integrating the validated model into clinical workflows and decision support systems to assist healthcare providers in patient care. It may also involve monitoring the model's performance over time and updating it with new data as needed.

Some examples of predictive modeling applications in healthcare include the following:

1. Risk prediction: Predictive models can be used to identify patients at high risk of developing chronic diseases, such as diabetes, heart disease, and cancer, based on their demographic, clinical, and genetic risk factors (Damen et al., 2016). This can help target preventive interventions and screenings to the most vulnerable populations.

2. Readmission prediction: Predictive models can estimate the likelihood of a patient being readmitted to the hospital within a certain time frame after discharge, based on factors such as their age, comorbidities, medication

use, and social determinants of health (Artetxe et al., 2018). This can help identify patients who may benefit from transitional care interventions and post-discharge follow-up.

3. Treatment response prediction: Predictive models can be used to estimate the likelihood of a patient responding to a particular treatment, such as a medication or surgical procedure, based on their individual characteristics and medical history (Obermeyer & Emanuel, 2016). This can help personalize treatment plans and avoid unnecessary or ineffective interventions.

4. Adverse event prediction: Predictive models can identify patients at high risk of experiencing adverse events, such as hospital-acquired infections, surgical complications, or medication side effects, based on their clinical and treatment-related factors (Goldstein et al., 2017). This can help implement targeted monitoring and prevention strategies to improve patient safety.

However, there are also challenges and limitations in developing and applying predictive models in healthcare, such as the need for large, high-quality, and representative datasets, the complexity and heterogeneity of patient populations, and the potential for bias and discrimination in model predictions (Char et al., 2018). Additionally, the interpretability and actionability of predictive models are critical for their adoption and trust by healthcare providers and patients (Tonekaboni et al., 2019).

Practical Tip: When developing predictive models for healthcare applications, it is important to involve domain experts, such as clinicians and epidemiologists, in the problem formulation, data selection, and model evaluation stages to ensure the clinical relevance and validity of the predictions (Goldstein et al., 2017). Using interpretable machine learning techniques, such as decision trees, rule-based models, and attention mechanisms, can help explain the reasoning behind the model predictions and facilitate their communication to end-users (Tonekaboni et al., 2019). Additionally, regularly updating and recalibrating the models with new data can help maintain their accuracy and generalizability over time (Davis et al., 2020).

CHAPTER 6

CORE AREAS OF CLINICAL INFORMATICS

6.1 CLINICAL INFORMATICS

Clinical informatics is a field that combines healthcare, information technology, and data analysis to improve patient care and streamline healthcare processes. It focuses on using digital tools and data-driven insights to support clinical decision-making, optimize workflows, and foster patient engagement. As healthcare systems worldwide face challenges such as increasing costs, demographic shifts, and growing complexity, clinical informatics plays a pivotal role in driving transformation and innovation.

The scope of clinical informatics is broad, encompassing various technologies, methodologies, and applications. Clinical informatics aims to harness data and technology to empower clinicians, administrators, and patients to make informed decisions and improve outcomes. This requires a deep understanding of both healthcare practices and the technical tools that can be applied to solve problems in the healthcare domain.

The rapid digitization of healthcare data has been a significant catalyst for the growth of clinical informatics. The adoption of electronic health records, the rise of mobile health applications and wearable devices, and the increasing availability of genomic and other advanced data types have created rich repositories of digital health information. These data assets can be analyzed to uncover insights and inform the development of innovative tools and interventions.

Concurrently, progress in AI, particularly in fields like machine learning, natural language processing, and computer vision, has revealed new opportunities for automating and enhancing clinical decision-making. AI algorithms can learn from vast datasets to identify patterns and generate predictions that assist clinicians in diagnosing diseases, tailoring treatments, and tracking patient progress.

However, integrating AI into clinical informatics also presents significant challenges and considerations. Technical hurdles related to data quality, system interoperability, and scalability must be addressed. Ethical and legal questions surrounding privacy, bias, and responsibility demand attention (Wiens, 2019). Ensuring the safety, effectiveness, and fairness of AI tools requires sustained collaboration among healthcare providers, technology developers, policymakers, and patients.

In the subsequent sections, we explore several core areas of clinical informatics in greater depth, including electronic health records, telemedicine, clinical decision support systems, and patient data management. We examine how AI is being applied in each of these domains and discuss the challenges and future directions for the field. By the conclusion of this chapter, readers will have gained a comprehensive understanding of the clinical informatics landscape and the transformative potential of AI in shaping the future of healthcare delivery.

6.2   THE EVOLUTION OF EHRS

Electronic health Records (EHRs) have become the digital foundation of modern healthcare, replacing traditional paper-based records in many settings. EHRs are digital repositories that store a wide array of patient health information, including demographics, medical history, medications, allergies, immunizations, laboratory results, and diagnostic images. The transition from paper to digital records has been driven by a recognition of the potential benefits of EHRs, as well as government initiatives and incentives.

In the United States, the Health Information Technology for Economic and Clinical Health (HITECH) Act of 2009 played a significant role in accelerating EHR adoption. The HITECH Act provided financial incentives for healthcare providers to adopt and meaningfully use certified EHR systems, aiming to improve the quality, safety, and efficiency of healthcare delivery. Similar initiatives have been undertaken in other countries to promote the digitization of health records.

The shift toward EHRs has brought numerous advantages to healthcare organizations and patients alike. EHRs enable more efficient storage, retrieval, and sharing of patient data, reducing the risk of lost or misplaced records. They facilitate better care coordination by allowing authorized healthcare providers to access a patient's complete medical history, regardless of where the care was received. EHRs also support the implementation of clinical decision support tools, which can alert providers to potential drug interactions, recommend evidence-based treatments, and help prevent medical errors.

EHRs have the potential to empower patients by providing them with secure access to their own health information. Many EHR systems now include patient portals, which allow individuals to view their medical records, communicate with their healthcare providers, and take a more active role in managing their health.

Despite these benefits, the adoption and use of EHRs have not been without challenges. Interoperability, or the ability of different EHR systems to exchange and use information effectively, remains a significant hurdle. Inconsistent data standards and proprietary systems can hinder the seamless sharing of patient data across healthcare organizations. Additionally, concerns about data privacy and security have grown as the volume of digital health information has increased, necessitating robust safeguards to protect sensitive patient information from unauthorized access or breaches.

Another challenge associated with EHRs is the potential for information overload and user fatigue. The sheer amount of data captured in EHRs can make it difficult for healthcare providers to quickly identify the most relevant information for a given patient or situation. Poorly designed user interfaces and cumbersome documentation requirements can contribute to burnout among clinicians, who may spend more time interacting with the EHR than engaging directly with patients.

As EHRs continue to evolve, there is a growing recognition of the need to optimize these systems to better support clinical workflows and decision-making. This is where AI and machine learning technologies are increasingly important. In the next section, we consider how AI is being used to enhance EHRs and unlock the full potential of digital health data.

6.2.1 AI-Driven EHR Optimization

While EHRs have brought significant benefits to healthcare, they have also introduced new challenges that can hinder their effectiveness and usability. These challenges include information overload, alert fatigue, and increased documentation burden on clinicians. AI has emerged as a promising solution to address these issues by making EHRs more intelligent, user-friendly, and efficient.

One application of AI in EHR optimization is natural language processing (NLP). NLP algorithms can analyze unstructured clinical notes, which often contain valuable patient information, and extract relevant data to populate structured fields in the EHR automatically. This reduces the need for manual data entry, saving clinicians time and effort while ensuring that important information is captured consistently. NLP can also be used to generate clinical summaries, highlight key findings, and support more efficient documentation practices.

AI can significantly impact EHR optimization through predictive analytics. Machine learning algorithms can analyze vast amounts of EHR data to identify patterns and predict patient outcomes, such as the risk of hospital readmission, adverse drug events, or the development of chronic conditions. By using these insights, healthcare providers can proactively intervene and provide targeted care to patients at higher risk, ultimately improving outcomes and reducing healthcare costs.

AI combats alert fatigue, a common problem associated with EHRs. When clinicians receive excessive or irrelevant alerts, they may start to ignore or override them, potentially missing critical information. AI algorithms can learn from user behavior and feedback to prioritize and personalize alerts, ensuring that clinicians receive the most relevant and actionable information at the right time. This can help reduce cognitive burden, enhance decision-making, and improve patient safety.

AI enables more intelligent and adaptive user interfaces within EHRs. By analyzing user behavior and preferences, AI algorithms can dynamically adjust the layout and content of EHR screens to highlight the most relevant information for each user and clinical context. This can help streamline workflows, reduce cognitive load, and improve the overall user experience for healthcare providers.

Despite the promising potential of AI in EHR optimization, there are also significant challenges and considerations to address. One challenge is ensuring the quality and integrity of the data used to train AI algorithms. EHR data can be incomplete, inconsistent, or biased, which can lead to inaccurate or misleading predictions. Robust data governance practices, including data standardization, validation, and cleansing, are essential to ensure the reliability and effectiveness of AI-driven EHR optimization.

Another important consideration is the need for transparency and explainability in AI algorithms. Healthcare providers must be able to understand and trust the recommendations and insights generated by AI systems. This requires the development of interpretable AI models that can provide clear explanations for their outputs and allow clinicians to review and validate the underlying logic. Ensuring the transparency and accountability of AI in EHRs is crucial for building trust and facilitating adoption among healthcare professionals.

Furthermore, the implementation of AI in EHRs raises important ethical and legal considerations. There are concerns about potential biases in AI algorithms, which could exacerbate existing health disparities if not carefully addressed. Ensuring the fairness, non-discrimination, and inclusivity of AI-driven EHR optimization is a critical challenge that requires ongoing attention and collaboration among healthcare providers, technology developers, and policymakers.

Looking to the future, the integration of AI in EHRs holds immense promise for transforming healthcare delivery. As AI technologies continue to advance and mature, we can expect to see more sophisticated and personalized EHR systems that adapt to the needs of individual users and clinical contexts. However, realizing this potential will require a concerted effort to address the technical, ethical, and regulatory challenges associated with AI in healthcare. By working together to develop robust, transparent, and equitable AI solutions, we can utilize the power of data and technology to optimize EHRs and ultimately improve patient care and outcomes.

6.2.2 Challenges and Future Directions

While the integration of AI in EHRs offers significant potential for optimization and improvement, it also presents a range of challenges that must be addressed to ensure successful implementation and adoption. These challenges span technical, operational, and ethical dimensions, requiring a collaborative and multidisciplinary approach to navigate effectively.

One of the primary technical challenges is data quality and standardization. EHRs often contain incomplete, inconsistent, or inaccurate data, which can hinder the effectiveness of AI algorithms. Ensuring that EHR data is clean, reliable, and properly formatted is essential for training accurate and trustworthy AI models. This requires the development and adoption of robust data governance frameworks, including standardized terminologies, data models, and quality assurance processes. Initiatives such as the Fast Healthcare Interoperability Resources (FHIR) standard aim to promote data interoperability and facilitate the exchange of health information across different systems and organizations.

Another significant challenge is the interoperability of EHR systems. Despite efforts to promote data exchange and collaboration, many EHR systems remain siloed and incompatible with one another. This lack of interoperability hinders the ability to aggregate and analyze data across multiple sources, limiting the potential insights and benefits of AI-driven optimization. Overcoming this challenge will require continued efforts to develop and implement open standards, APIs, and data sharing protocols that enable seamless communication and integration between different EHR systems.

Privacy and security concerns also pose significant challenges to the adoption of AI in EHRs. As AI algorithms require access to large volumes of patient data, ensuring the confidentiality and protection of sensitive health information is paramount. Robust data security measures, such as encryption, access controls, and audit trails, must be implemented to safeguard patient privacy and prevent unauthorized access or misuse of data. Additionally, clear policies and guidelines around data ownership, consent, and governance are necessary to maintain patient trust and comply with regulatory requirements, such as the Health Insurance Portability and Accountability Act (HIPAA) in the United States and the General Data Protection Regulation (GDPR) in the European Union.

From an ethical perspective, the use of AI in EHRs raises important questions about bias, fairness, and accountability. AI algorithms can inadvertently perpetuate or amplify existing biases in healthcare data, leading to disparities in treatment recommendations or resource allocation. Ensuring that AI models are developed and validated using diverse and representative datasets, and incorporating principles of fairness and non-discrimination, is crucial to mitigate these risks. Establishing clear mechanisms for transparency, explainability, and human oversight of AI-driven decisions is essential to maintain accountability and build trust among healthcare providers and patients.

Looking to the future, the continued evolution of AI in EHRs holds immense promise for transforming healthcare delivery. As AI technologies advance, we can expect to see more sophisticated and adaptive EHR systems that leverage real-time data streams, personalized recommendations, and intelligent automation to support clinical decision-making and optimize workflows. The integration of AI with other emerging technologies, such as blockchain for secure data sharing and Internet of Things (IoT) devices for continuous patient monitoring, could further enhance the capabilities and impact of EHRs.

However, realizing this potential will require ongoing collaboration and investment from healthcare organizations, technology vendors, policymakers, and researchers. Continued research and evaluation are needed to validate the effectiveness and safety of AI interventions in real-world clinical settings, and to identify and address any unintended consequences or limitations. Engaging healthcare providers and patients in the design, development, and deployment of AI-driven EHR solutions will be essential to ensure their usability, acceptability, and ultimate success.

While the challenges associated with AI in EHRs are significant, so too are the opportunities for transforming healthcare delivery and improving patient outcomes. By working together to address the technical, ethical, and operational hurdles, and by fostering a culture of innovation, collaboration, and continuous learning, we can harness the power of AI to optimize EHRs and shape a future of more intelligent, efficient, and patient-centered care.

Practical Tips

1. Data governance framework: To ensure data quality and standardization, healthcare organizations should implement clear policies and procedures for data collection, validation, and management. This includes adopting standardized terminologies, such as SNOMED CT and LOINC, and regularly auditing EHR data for completeness and accuracy. Investing in data cleansing and normalization tools can also help improve data quality and consistency.

2. Prioritize interoperability: When selecting or upgrading EHR systems, prioritize those that adhere to open standards and APIs, such as FHIR, to facilitate data exchange and integration. Engage with industry consortia and initiatives that promote interoperability, such as the Argonaut Project and the Sequoia Project, to stay informed about best practices and emerging solutions.

3. Privacy and security measures: Ensure that your organization has robust data security policies and technologies in place, including encryption, access controls, and audit trails. Regularly train staff on data privacy and security best practices, and conduct periodic risk assessments to identify and address potential vulnerabilities. Stay up-to-date with evolving

regulatory requirements, such as HIPAA and GDPR, and ensure compliance through regular audits and updates.

4. Bias and fairness in AI models: When developing or deploying AI algorithms for EHRs, ensure that the training data is diverse, representative, and free from historical biases. Implement fairness metrics and auditing processes to detect and mitigate potential biases in AI-driven recommendations or decisions. Engage with diverse stakeholders, including patients and community representatives, to gather feedback and ensure that AI solutions are inclusive and equitable.

5. Transparency and explainability: Develop AI models that provide clear explanations for their outputs and decision-making processes. Use intuitive visualizations and user interfaces to help clinicians understand and interpret AI-driven insights. Establish governance mechanisms for human oversight and intervention, allowing clinicians to review and override AI recommendations when necessary. Regularly communicate with patients about the use of AI in their care and provide opportunities for informed consent and feedback.

6. Research and evaluation: Allocate resources for ongoing research and evaluation of AI-driven EHR interventions. Collaborate with academic institutions and research organizations to conduct rigorous studies on the effectiveness, safety, and impact of AI solutions in real-world clinical settings. Share findings and best practices with the broader healthcare community to accelerate learning and adoption.

7. Innovation and collaboration: Foster a culture that encourages experimentation, learning, and continuous improvement. Provide opportunities for cross-disciplinary collaboration among clinicians, data scientists, and IT professionals to co-design and iterate on AI solutions. Engage with patients and caregivers to gather insights and feedback on their needs and preferences. Participate in industry forums and conferences to share experiences and learn from peers.

By implementing these practical tips and considerations, healthcare organizations can more effectively address the challenges and opportunities associated with AI in EHRs. As the field continues to evolve, it will be essential to remain adaptable, proactive, and committed to realizing the full potential of AI-driven EHR optimization for improving patient care and outcomes.

6.3 THE RISE OF TELEMEDICINE

Telemedicine, the use of electronic communication and information technologies to provide remote clinical healthcare services, has experienced significant growth in recent years. This growth has been driven by advancements in technology, changes in healthcare policies and reimbursement models, and an increasing demand for convenient and accessible care.

The concept of telemedicine is not new, with early examples dating back to the 1960s when NASA used telemetry to monitor the health of astronauts in space. However, the widespread adoption of telemedicine has been relatively recent, accelerated by the proliferation of high-speed internet, smartphones, and other digital technologies.

Telemedicine encompasses a wide range of services, including the following:

1. Video consultations: Patients can connect with healthcare providers through secure video conferencing platforms, enabling real-time, face-to-face interactions without the need for in-person visits.

2. Remote monitoring: Patients with chronic conditions or those recovering from acute illnesses can use connected devices, such as wearable sensors and smart health monitors, to transmit vital signs and other health data to their providers for continuous monitoring and early intervention.

3. Asynchronous communication: Patients can securely message their providers, share images or videos, and receive feedback and guidance through patient portals or dedicated telemedicine platforms.

4. Tele-triage: Patients can access nurse triage lines or chatbots to receive initial assessments, guidance on self-care, and recommendations for appropriate levels of care based on their symptoms and needs.

The benefits of telemedicine are numerous. For patients, telemedicine offers increased convenience, accessibility, and flexibility. It eliminates the need for travel and time off work, reduces exposure to infectious diseases in waiting rooms, and enables patients in rural or underserved areas to access specialist care. For healthcare providers, telemedicine can improve efficiency, reduce no-shows, and expand their reach beyond traditional geographic boundaries.

Telemedicine has the potential to improve health outcomes and reduce healthcare costs. By enabling early intervention, continuous monitoring, and patient engagement, telemedicine can help prevent complications, reduce hospital readmissions, and promote better disease management.

The COVID-19 pandemic has further accelerated the adoption of telemedicine, as healthcare systems worldwide sought to minimize in-person contact and reduce the risk of virus transmission. Many regulatory barriers and reimbursement restrictions were temporarily lifted to facilitate the rapid expansion of telemedicine services. This has led to a significant increase in the use of telemedicine across various specialties and settings, from primary care to mental health services.

Practical Tip: Healthcare providers looking to implement or expand telemedicine services should carefully evaluate the available technology platforms, ensuring that they are secure, user-friendly, and compliant with relevant regulations, such as HIPAA. Providers should also invest in training and support for staff and patients to ensure smooth adoption and optimal use of telemedicine tools.

As telemedicine continues to evolve and mature, it will become an integral part of healthcare delivery. However, realizing its full potential will require ongoing efforts to address challenges related to reimbursement, licensure, quality assurance, and digital literacy. In the next section, we explore how AI is being used to further enhance the capabilities and impact of telemedicine.

6.3.1 AI in Remote Patient Monitoring

Remote patient monitoring (RPM) is an application of telemedicine that involves the use of connected devices and sensors to collect and transmit patient health data from outside traditional healthcare settings. AI algorithms can analyze this data in real-time, enabling early detection of deterioration, prediction of adverse events, and personalized interventions. The integration of AI in RPM has the potential to revolutionize the management of chronic conditions, post-acute care, and population health.

One of the most promising use cases for AI in RPM is the early detection and prediction of clinical deterioration. By continuously analyzing patient vital signs, such as heart rate, respiratory rate, and blood pressure, AI algorithms can identify subtle changes and patterns that may indicate the onset of complications or exacerbations. For example, machine learning models can be trained to detect early signs of sepsis, a life-threatening condition that requires prompt intervention. By alerting healthcare providers to these early warning signs, AI-enabled RPM can facilitate timely interventions and improve patient outcomes.

Another area where AI can add significant value to RPM is in the management of chronic conditions, such as diabetes, heart failure, and chronic obstructive pulmonary disease (COPD). AI algorithms can analyze data from wearable devices, smart health monitors, and patient-reported outcomes to provide personalized insights and recommendations. For instance, an AI-powered glucose monitoring system can learn an individual's glucose patterns and provide real-time guidance on insulin dosing, diet, and exercise. Similarly, AI can analyze data from smart inhalers to assess medication adherence and technique, and provide targeted education and feedback to patients with asthma or COPD.

AI can also enhance the efficiency and scalability of RPM by automating certain tasks and workflows. Natural language processing (NLP) can be used to analyze unstructured data, such as patient-reported symptoms or free-text notes, and extract relevant information for clinical decision-making. Computer vision algorithms can analyze images or videos captured by patients, such as wound photos or gait recordings, and provide objective assessments and recommendations. These AI-driven automations can help reduce the burden on healthcare providers and enable them to focus on higher-value tasks and patient interactions.

Practical Tip: When implementing AI in RPM, it is crucial to ensure the quality and reliability of the data being collected. This requires the use of validated and calibrated devices, clear patient instructions, and robust data transmission and storage protocols. Healthcare providers should also establish clear guidelines for responding to AI-generated alerts and recommendations, ensuring that there is human oversight and clinical judgment involved in decision-making.

Despite the potential benefits of AI in RPM, there are also important challenges and considerations to address. One key challenge is the integration of AI-generated insights into clinical workflows and decision-making processes. Healthcare providers need to be trained on how to interpret and act upon AI recommendations, and there must be clear protocols for escalation and human intervention when necessary. Another challenge is ensuring the privacy and security of patient data, particularly as it is being transmitted and analyzed by AI algorithms. Robust data governance frameworks and encryption protocols are essential to protect patient confidentiality and prevent unauthorized access or misuse of data.

There are concerns about the potential for bias and disparities in AI-driven RPM. If AI algorithms are trained on data that is not representative of diverse patient populations, they may generate recommendations that are not applicable or appropriate for certain groups. Ensuring that AI models are developed and validated using diverse and inclusive datasets, and incorporating principles of fairness and equity, is crucial to mitigate these risks.

The integration of AI in RPM has the potential to transform the delivery of care for patients outside traditional healthcare settings. By enabling early detection, personalized recommendations, and automated workflows, AI can improve the efficiency, effectiveness, and scalability of RPM. However, realizing this potential will require ongoing collaboration and innovation among healthcare providers, technology vendors, and policymakers to address the technical, ethical, and operational challenges associated with AI in RPM.

6.3.2 Overcoming Barriers to Adoption

Despite the numerous potential benefits of AI-enabled telemedicine and remote patient monitoring (RPM), there are several significant barriers that must be addressed to facilitate widespread adoption and implementation. These barriers span technical, regulatory, financial, and cultural domains, requiring a multifaceted and collaborative approach to overcome.

One of the primary technical barriers is the lack of interoperability and standardization among telemedicine and RPM systems. Many devices, platforms, and data formats are proprietary and incompatible with one another, hindering the ability to seamlessly exchange and analyze data across different settings and providers. The development and adoption of open standards, such as FHIR (Fast Healthcare Interoperability Resources), can help promote interoperability and facilitate the integration of AI-driven insights into clinical workflows.

Another technical challenge is ensuring the reliability and accuracy of data collected through telemedicine and RPM. Poor data quality, due to factors such as user error, device malfunctions, or network disruptions, can compromise the effectiveness of AI algorithms and lead to incorrect or misleading recommendations. Implementing robust data validation, cleansing, and preprocessing techniques, as well as providing clear patient and provider education on proper device use and data collection protocols, can help mitigate these risks.

From a regulatory perspective, telemedicine and RPM are subject to a complex and evolving landscape of laws, regulations, and policies that vary by jurisdiction and payer. Issues such as licensure, credentialing, privacy, and security must be carefully navigated to ensure compliance and protect patient rights. The lack of clear and consistent guidelines can create uncertainty and hesitation among healthcare organizations looking to implement AI-driven telemedicine solutions. Collaboration among policymakers, regulators, and industry stakeholders is needed to develop coherent and supportive frameworks that balance innovation with patient safety and privacy.

Financial barriers, such as reimbursement models and investment costs, also pose significant challenges to the adoption of AI in telemedicine. Traditional fee-for-service models may not adequately compensate providers for telemedicine and RPM services, particularly those that involve AI-driven decision support or care management. The development of value-based reimbursement models, such as bundled payments or capitation, can help align incentives and support the long-term sustainability of AI-driven telemedicine programs. The upfront costs of acquiring and implementing AI technologies can be substantial, particularly for smaller healthcare organizations with limited resources. Innovative financing mechanisms, such as risk-sharing agreements or pay-for-performance contracts, can help mitigate these costs and ensure a positive return on investment.

Cultural and organizational barriers, such as resistance to change and lack of digital literacy, can also hinder the adoption of AI in telemedicine. Healthcare providers may be hesitant to trust or rely on AI-generated recommendations, particularly if they are not transparent or explainable. Providing education and training on the capabilities and limitations of AI, as well as involving providers in the design and validation of AI models, can help build trust and acceptance. Patients may have varying levels of comfort and familiarity with telemedicine and RPM technologies, particularly among older or underserved populations. Providing patient education, support, and resources, such as multilingual materials or community health workers, can help bridge these digital divides and ensure equitable access to AI-driven telemedicine services.

Practical Tip: To overcome barriers to adoption, healthcare organizations should take a phased and iterative approach to implementing AI in telemedicine. Starting with small-scale pilot projects, gathering feedback from providers and patients, and continuously refining and improving the solutions can help build momentum and demonstrate value. Collaborating with experienced technology vendors, academic institutions, and other healthcare organizations can provide valuable expertise, resources, and best practices to guide the implementation process.

Another significant barrier to the adoption of AI in telemedicine is the physician-centric nature of many healthcare models. Traditionally, healthcare has been structured around the primacy of physician decision-making, with other healthcare professionals and technologies playing a supportive role. This physician-centric approach can create resistance to the integration of AI-driven insights and recommendations, particularly if they are perceived as challenging or superseding physician judgment.

Physicians may be hesitant to rely on AI algorithms for several reasons. They may question the accuracy, reliability, and applicability of AI-generated recommendations, particularly if they are based on data or populations that differ from their own practice. They may also be concerned about the potential for AI to disrupt established workflows, reduce autonomy, or introduce new liabilities. The "black box" nature of some AI models, where the underlying logic and reasoning are not easily interpretable, can further erode physician trust and acceptance.

To overcome this barrier, it is essential to engage physicians as stakeholders in the development, validation, and implementation of AI solutions in telemedicine. This involves providing education and training on the capabilities and limitations of AI, as well as the evidence base supporting its use. It also requires involving physicians in the design and iteration of AI models, ensuring that they are transparent, explainable, and aligned with clinical best practices. By fostering a collaborative and inclusive approach, healthcare organizations can help physicians view AI as a valuable tool to augment, rather than replace, their clinical judgment and decision-making.

The integration of AI in telemedicine should be framed as enhancing clinician-patient relationships and improve patient outcomes, rather than a threat to physician autonomy or authority. AI-driven insights can help physicians gain a more comprehensive understanding of their patients' health status, identify potential risks or opportunities for intervention, and tailor treatment plans to individual needs and preferences. By positioning AI as a tool to support shared decision-making and patient-centered care, healthcare organizations can help align physician incentives and motivations with the adoption of these technologies.

Practical Tip: To address physician concerns and promote adoption, healthcare organizations should establish clear governance structures and protocols for the use of AI in telemedicine. This includes defining the roles and responsibilities of physicians, clinicians, data scientists, and other stakeholders, as well as establishing processes for reviewing, validating, and updating AI models over time. Regular communication and feedback loops between physicians and AI teams can help ensure that the solutions remain clinically relevant, actionable, and trustworthy.

The physician-centric nature of healthcare models poses a significant barrier to the adoption of AI in telemedicine. By engaging physicians and clinicians as stakeholders, providing education and training, and framing AI as a tool to support patient-centered care, healthcare organizations can help overcome this barrier and realize the full potential of these technologies to improve healthcare delivery and outcomes. As the field continues to evolve, it will be essential to maintain an ongoing dialogue and collaboration between physicians, data scientists, and other stakeholders to ensure that AI solutions are developed and implemented in a way that is clinically meaningful, ethically sound, and patient-focused.

While the barriers to adopting AI in telemedicine are significant, they are not insurmountable. By addressing technical, regulatory, financial, and cultural challenges through collaboration, education, and innovation, healthcare organizations can unlock the full potential of AI-driven telemedicine to improve access, quality, and efficiency of care. As the field continues to evolve and mature, it will be essential to remain vigilant and proactive in identifying and addressing new barriers and opportunities as they emerge.

6.4 THE ROLE OF CDSSS IN PATIENT CARE

Clinical Decision Support Systems (CDSSs) play a crucial role in enhancing patient care by providing healthcare professionals with timely, relevant, and evidence-based information to support clinical decision-making. CDSSs are computer-based tools that leverage vast amounts of clinical data, medical knowledge, and patient-specific information to generate actionable insights and recommendations at the point of care.

The primary goal of CDSSs is to improve the quality, safety, and efficiency of healthcare delivery by assisting clinicians in making informed decisions. This is achieved through various mechanisms, such as

1. Alerts and reminders: CDSSs can generate real-time alerts and reminders to help clinicians identify potential drug interactions, contraindications, or missed preventive care opportunities. For example, a CDSS might alert a physician about a patient's allergies when prescribing a new medication or

remind them to order a mammogram for a patient due for breast cancer screening.

2. Diagnostic support: CDSSs can assist clinicians in the diagnostic process by suggesting potential differential diagnoses based on a patient's symptoms, medical history, and test results. These systems can also provide links to relevant medical literature, clinical guidelines, and diagnostic algorithms to support clinical reasoning and decision-making.

3. Treatment recommendations: CDSSs can provide personalized treatment recommendations based on a patient's specific characteristics, such as age, gender, comorbidities, and genetic profile. These recommendations can be based on clinical practice guidelines, expert consensus, or machine learning algorithms that analyze large datasets to identify optimal treatment strategies.

4. Risk stratification: CDSSs can help clinicians identify patients at high risk for adverse outcomes, such as hospital readmissions, complications, or disease progression. By utilizing predictive analytics and machine learning techniques, CDSSs can analyze patient data to generate risk scores and stratify patients into different risk categories, enabling targeted interventions and resource allocation.

5. Quality improvement: CDSSs can support quality improvement initiatives by monitoring clinical performance metrics, identifying gaps in care, and suggesting evidence-based interventions to address them. For example, a CDSS might track a healthcare organization's compliance with sepsis management protocols and provide feedback and recommendations to improve adherence and outcomes.

The integration of CDSSs into clinical workflows has the potential to significantly improve patient outcomes, reduce medical errors, and enhance healthcare efficiency. Studies have shown that CDSSs can increase adherence to clinical guidelines, reduce adverse drug events, and improve the appropriateness of diagnostic testing and treatment selection.

However, the effectiveness of a CDSS depends on several factors, including the quality and relevance of the underlying clinical knowledge base, the usability and integration of the system into clinical workflows, and the acceptance and adoption by healthcare professionals. Poorly designed or implemented CDSS can lead to alert fatigue, workflow disruptions, and unintended consequences, such as overreliance on technology or decreased critical thinking skills.

Practical Tip: When implementing a CDSS, healthcare organizations should involve end-users, such as physicians and nurses, in the design and testing process to ensure that the system is user-friendly, clinically relevant, and aligned with existing workflows. Providing adequate training and support, as well as establishing clear protocols for managing and updating the clinical knowledge base, can help ensure the long-term success and sustainability of CDSS initiatives.

CDSSs play a vital role in supporting clinical decision-making and improving patient care. By utilizing advanced analytics, machine learning, and clinical knowledge, CDSSs can provide healthcare professionals with actionable insights and recommendations to optimize diagnosis, treatment, and care management. As CDSSs change, it will be essential to address challenges related to data quality, interoperability, and user acceptance to fully realize the potential of these powerful tools to transform healthcare delivery.

6.4.1 AI Techniques in CDSSs

While Chapter 2 provided an overview of foundational AI technologies, such as machine learning and natural language processing, this section discusses specific AI techniques and applications that are transforming the capabilities and performance of Clinical Decision Support Systems (CDSSs).

One promising application of AI in CDSSs is the use of deep learning models for analyzing complex clinical data. Deep learning architectures, such as convolutional neural networks (CNNs) and recurrent neural networks (RNNs), can automatically learn hierarchical representations of data, enabling them to capture subtle patterns and dependencies that may be difficult for traditional ML algorithms to detect (Vaswani et al., 2017). For example, CNNs have been used to analyze medical images, such as chest X-rays and retinal scans, to detect signs of disease with high accuracy. RNNs are well-suited for analyzing time-series data, such as electronic health records and physiological signals, to predict patient trajectories and identify early warning signs of deterioration.

Another important AI technique in CDSSs is knowledge representation and reasoning. This involves encoding clinical knowledge, such as clinical guidelines, expert rules, and disease ontologies, into machine-readable formats that can be used by CDSS to generate recommendations and support decision-making. One promising approach is the use of *knowledge graphs*, which represent clinical knowledge as a network of entities and relationships.

By utilizing techniques from graph theory and semantic Web technologies, knowledge graphs can enable more flexible and context-aware reasoning in CDSS. For example, a knowledge graph could be used to identify potential drug-drug interactions based on the molecular pathways and pharmacological properties of the drugs, rather than relying on predefined rules or blacklists.

Hybrid AI approaches, which combine multiple AI techniques, are also gaining traction in CDSS. For example, a hybrid system might use machine learning to analyze patient data and generate initial hypotheses, and then use knowledge-based reasoning to refine and validate the hypotheses based on clinical guidelines and expert knowledge. This method can help overcome the limitations of purely data-driven or knowledge-based approaches and provide more robust and explainable recommendations.

Practical considerations for implementing AI in CDSSs include the following:

1. Data quality and standardization: Ensuring that the clinical data used to train and validate AI models is accurate, complete, and consistently formatted is critical for the success of AI-powered CDSS. This may require data cleaning, harmonization, and normalization techniques, as well as the use of standardized terminologies and data models.

2. Model interpretability and transparency: To build trust and acceptance among clinicians, it is important to ensure that the AI models used in CDSS are interpretable and transparent. This may involve using techniques such as feature importance analysis, model visualization, and natural language explanations to provide insights into how the models are making decisions.

3. Continuous learning and adaptation: As clinical knowledge and patient populations change, it is important to ensure that the AI models used in CDSS are continuously updated and adapted to reflect the latest evidence and best practices. This may involve techniques such as online learning, transfer learning, and active learning to enable the models to learn from new data and feedback.

4. Human-AI collaboration: Rather than viewing AI as a replacement for human expertise, it is important to design CDSSs that support and augment human decision-making. This may involve techniques such as shared decision-making, explainable AI, and human-in-the-loop learning to enable clinicians and AI systems to work together effectively.

By utilizing these advanced AI techniques and addressing these practical considerations, CDSS can provide more accurate, personalized, and context-aware recommendations to support clinical decision-making and improve patient outcomes.

Code Tips

Introduction to Convolutional Neural Networks for Medical Image Analysis

Convolutional Neural Networks (CNNs) have revolutionized the field of medical imaging by providing powerful tools for automated analysis and classification of medical images. This guide presents a comprehensive workflow for developing a CNN model specifically designed for chest X-ray image classification, which can assist healthcare professionals in detecting abnormalities with greater accuracy and efficiency. The following code snippets and explanations walk through the essential steps of building, training, and evaluating a CNN model for binary classification of medical images, from data preprocessing and augmentation to model architecture design and performance evaluation.

Importing Required Libraries

```
import tensorflow as tf
from tensorflow import keras
```

Explanation

- tensorflow: TensorFlow is an open-source deep learning framework for building and training neural networks.
- keras: A high-level API within TensorFlow, simplifying the development of deep learning models.

Loading and Preprocessing the Chest X-Ray Images

This section describes how to properly load medical images from a structured directory and prepare them for training a CNN model. The preprocessing steps are critical for ensuring that all images are standardized to the same dimensions and properly labeled for classification.

```
train_images = keras.preprocessing.
image_dataset_from_directory(
    "path/to/train/images",
    label_mode='binary',
    image_size=(224, 224),
    batch_size=32
)

test_images = keras.preprocessing.
image_dataset_from_directory(
    "path/to/test/images",
    label_mode='binary',
    image_size=(224, 224),
    batch_size=32
)
```

Explanation

- image_dataset_from_directory:
  - loads images from a directory structure where subdirectories represent class labels
  - example directory structure:
    - path/to/train/images/
    - ├── Normal/
    - └── Abnormal/
- Parameters:
  - label_mode='binary': Indicates binary classification (e.g., normal vs. abnormal). Labels are assigned as 0 or 1.
  - image_size=(224, 224): Resizes all images to 224x224 pixels (common input size for CNNs).
  - batch_size=32: Groups 32 images into a batch during training/testing.

Enhancing Preprocessing with Data Augmentation

To improve model generalization, apply data augmentation to the training data.

```
data_augmentation = keras.Sequential([
    keras.layers.RandomFlip("horizontal"),
    keras.layers.RandomRotation(0.1),
    keras.layers.RandomZoom(0.1),
])


# Apply augmentation to the training dataset
train_images = train_images.map(lambda x, y: (data_augmentation(x), y))
```

Explanation

- data augmentation:
  - randomly flips, rotates, or zooms images during training, simulating real-world variability
  - helps prevent overfitting by increasing the diversity of training data
- map:
  - applies the data_augmentation transformations to the training images while preserving the labels

Defining the CNN Architecture

```
model = keras.Sequential([
    keras.layers.Conv2D(32, (3, 3), activation='relu', input_shape=(224, 224, 3)),
    keras.layers.MaxPooling2D((2, 2)),
    keras.layers.Conv2D(64, (3, 3), activation='relu'),
```

```
    keras.layers.MaxPooling2D((2, 2)),
    keras.layers.Conv2D(64, (3, 3),
activation='relu'),
    keras.layers.Flatten(),
    keras.layers.Dense(64, activation='relu'),
    keras.layers.Dense(1, activation='sigmoid')
])
```

Explanation

- keras.Sequential:
  - defines a linear stack of layers for the CNN
- Layers:
  0. Conv2D:
     - Convolutional layers extract features from images using filters (kernels).
     - 32, 64: number of filters in each layer
     - (3, 3): filter size (3x3 kernel)
     - activation='relu': Rectified Linear Unit (ReLU) activation introduces nonlinearity.
  1. MaxPooling2D:
     - downsamples feature maps by retaining the most dominant features
     - (2, 2): pooling size (reduces dimensions by half)
  2. Flatten:
     - converts the 2D feature maps into a 1D vector for the dense (fully connected) layers
  3. Dense:
     - fully connected layers for classification

- 64: hidden layer with 64 neurons
- 1: output layer with 1 neuron for binary classification

    activation='sigmoid': outputs probabilities for binary classification (0 or 1)

Compiling the Model

```
model.compile(
    optimizer='adam',
    loss='binary_crossentropy',
    metrics=['accuracy']
)
```

Explanation

- optimizer='adam':
  - Adaptive Moment Estimation (Adam) optimizer adjusts learning rates dynamically for efficient training.
- loss='binary_crossentropy':
  - loss function for binary classification problems
  - measures the difference between predicted and true labels
- metrics=['accuracy']:
  - tracks the accuracy during training and evaluation

Training the Model

After defining and compiling the CNN architecture, the next critical step is training the model on our prepared dataset. This section explains how to execute the training process and monitor performance.

```
history = model.fit(
    train_images,
    epochs=10,
    validation_data=test_images
)
```

Explanation:

- **model.fit**: This is the Keras function that trains the model on the provided dataset
- **train_images**: The preprocessed training dataset loaded in the previous step
- **epochs=10**: The number of complete passes through the entire training dataset
- **validation_data=test_images**: The separate test dataset used to evaluate the model after each epoch

During training, the model adjusts its weights to minimize the loss function (binary cross-entropy) that was specified during compilation. The training process generates a history object that tracks various metrics like accuracy and loss for both training and validation sets, which will be useful for performance visualization and analysis in later steps.

Retry

Claude can make mistakes.

Please double-check responses.

Add a callback to stop training early if the model stops improving.

```
early_stopping = keras.callbacks.EarlyStopping(
    monitor='val_loss', patience=3,
restore_best_weights=True
)
```

```
history = model.fit(
    train_images,
    epochs=10,
    validation_data=test_images,
    callbacks=[early_stopping]
)
```

Evaluating the Model

After training the CNN model on the chest X-ray images, it's essential to rigorously evaluate its performance on unseen data to assess how well it will generalize to new medical images in real-world scenarios.

```
test_loss, test_acc = model.evaluate(test_images)
print(f'Test accuracy: {test_acc:.3f}')
```

Explanation:

- **model.evaluate**:
  - Computes the loss and accuracy on the test dataset
- **Output**:
  - Displays the test accuracy, providing insight into how well the model generalizes to unseen data

This evaluation step is crucial for determining if the model is ready for deployment in clinical settings. A high test accuracy indicates that the model has successfully learned to distinguish between normal and abnormal chest X-rays. However, in medical applications, additional metrics such as sensitivity, specificity, and area under the ROC curve should also be considered for a more comprehensive evaluation of model performance.

Visualizing Training Performance

To better understand model performance, plot the training and validation accuracy/loss over epochs.

```python
import matplotlib.pyplot as plt

# Plot training and validation accuracy
plt.plot(history.history['accuracy'], label='Training Accuracy')
plt.plot(history.history['val_accuracy'], label='Validation Accuracy')
plt.xlabel('Epochs')
plt.ylabel('Accuracy')
plt.title('Training and Validation Accuracy')
plt.legend()
plt.show()


# Plot training and validation loss
plt.plot(history.history['loss'], label='Training Loss')
plt.plot(history.history['val_loss'], label='Validation Loss')
plt.xlabel('Epochs')
plt.ylabel('Loss')
plt.title('Training and Validation Loss')
plt.legend()
plt.show()
```

Explanation

- history.history:
  - stores the training and validation accuracy/loss for each epoch
- Visualization:
  - helps identify overfitting (when validation accuracy diverges from training accuracy)

Saving the Model

```
model.save('cnn_chest_xray_model.h5')
```

Explanation:

- model.save: saves the model architecture, weights, and optimizer configuration to a file (.h5 format)

TABLE 6.1 Summary of Code Workflows

Step	Purpose
Data Loading	Loads and preprocesses chest X-ray images from directories.
Data Augmentation	Augments training data to improve generalization.
CNN Architecture	Defines a convolutional neural network for image classification.
Compilation	Configures the optimizer, loss function, and evaluation metrics.
Training	Trains the model on the training dataset and monitors validation metrics.
Evaluation	Evaluates the model's performance on the test dataset.
Visualization	Plots training/validation accuracy and loss trends over epochs.
Saving	Saves the trained model for deployment or further analysis.

Knowledge Graph for Drug-Drug Interaction Analysis: Importing the Necessary Library

Drug-drug interactions (DDIs) represent a significant clinical challenge that can lead to adverse events, reduced therapeutic efficacy, or increased toxicity. Knowledge graphs provide a powerful framework for modeling these complex interactions by representing drugs, their properties, and their relationships in an intuitive network structure.

Importing the Necessary Library

```
import networkx as nx
```

Explanation:

- **networkx**: A Python library for creating, manipulating, and analyzing graphs and networks. Here, we use it to build a knowledge graph for drug-drug interactions (DDIs).

With this library, we can construct a comprehensive knowledge graph where nodes represent drugs and edges represent various types of interactions between them. This graph-based approach enables advanced analysis techniques such as path finding to discover potential indirect interactions, centrality measures to identify critical drugs in the interaction network, and community detection to identify clusters of frequently interacting medications.

Retry

Claude can make mistakes.

Please double-check responses.

Creating the Knowledge Graph

```
# Create a directed graph
G = nx.DiGraph()


# Add drugs and their properties as nodes
G.add_node("Drug1", type="drug", mechanism="inhibitor", target="Enzyme1")
G.add_node("Drug2", type="drug", mechanism="substrate", target="Enzyme1")
G.add_node("Drug3", type="drug", mechanism="inducer", target="Enzyme2")


# Add interactions as edges
G.add_edge("Drug1", "Drug2", type="interaction", effect="increased_exposure")
G.add_edge("Drug3", "Drug1", type="interaction", effect="decreased_exposure")
```

Explanation:

1. **Graph type:**
   - nx.DiGraph(): creates a directed graph where edges have a direction (e.g., Drug1 → Drug2)
2. **Nodes:**
   - represent drugs with associated metadata:
     - type="drug": specifies the node type
     - mechanism: indicates the drug's mechanism (e.g., inhibitor, substrate, inducer)
     - target: the biological target the drug interacts with (e.g., enzymes)
3. **Edges:**
   - Represent interactions between drugs:
     - type="interaction": classifies the edge as an interaction
     - effect: describes the effect of the interaction (e.g., increased or decreased exposure)

Querying the Knowledge Graph

After constructing a knowledge graph of drug interactions, we need efficient ways to query this network to identify potential interactions between medications. This functionality is especially valuable for clinical decision support systems that can alert healthcare providers about possible adverse interactions in a patient's medication regimen.

```
# Function to check for interactions between two drugs
def check_interaction(drug1, drug2):
    if nx.has_path(G, drug1, drug2):  # Check if a path exists between the two drugs
        path = nx.shortest_path(G, drug1, drug2)  # Get the shortest path
        interaction = G.get_edge_data(path[0], path[1])  # Retrieve edge data
```

```
        return f"Potential interaction: {drug1} may
lead to {interaction['effect']} of {drug2}"
    else:
        return f"No direct interaction found between
{drug1} and {drug2}"

# Example queries
print(check_interaction("Drug1", "Drug2"))
print(check_interaction("Drug1", "Drug3"))
```

Explanation:

1. **nx.has_path**:
   - Checks if there is a directed path from drug1 to drug2
2. **nx.shortest_path**:
   - Retrieves the shortest path (sequence of nodes) between the drugs
3. **G.get_edge_data**:
   - Extracts metadata (e.g., interaction type, effect) of the edge between the first two nodes in the path
4. **Query Results**:
   - If a path exists, the function returns the interaction and its effect.
   - If no path exists, it states that no direct interaction is found.

This querying mechanism allows healthcare professionals to quickly assess potential drug interactions, including both direct interactions and those that might occur through intermediary compounds, helping to prevent adverse drug events and improve patient safety.

Retry

Claude can make mistakes.

Please double-check responses.

Enhancing the Knowledge Graph

We can enhance the graph by adding additional drug properties, interactions, and visualization.

Adding More Nodes and Interactions

As our understanding of drug interactions evolves, it's essential to continuously update the knowledge graph with new drugs and newly discovered interactions. This section demonstrates how to expand the graph to incorporate additional medications and their relationship to existing drugs in the network.

```
# Add additional drugs and interactions
G.add_node("Drug4", type="drug",
mechanism="substrate", target="Enzyme2")
G.add_edge("Drug3", "Drug4", type="interaction",
effect="increased_exposure")
G.add_edge("Drug2", "Drug4", type="interaction",
effect="decreased_effect")
```

Explanation:

- **G.add_node**: Adds a new drug node to the graph with specified attributes:
  - **type**: Categorizes the node as a drug
  - **mechanism**: Describes how the drug interacts with biological systems (as a substrate)
  - **target**: Identifies the biological target of the drug (Enzyme2)
- **G.add_edge**: Creates connections between the new drug and existing drugs:
  - First edge: Indicates Drug3 causes increased exposure to Drug4
  - Second edge: Indicates Drug2 causes decreased effect of Drug4

This dynamic approach allows healthcare providers and researchers to maintain an up-to-date representation of the complex web of drug interactions, facilitating more accurate risk assessment and medication management for patients with multiple prescriptions.

Retry

Claude can make mistakes.

Please double-check responses.

Query Enhancement

Modify the function to display the entire interaction path, including intermediate drugs.

```
def check_interaction(drug1, drug2):
    if nx.has_path(G, drug1, drug2):
        path = nx.shortest_path(G, drug1, drug2)
        interaction_details = []
        for i in range(len(path) - 1):
            edge_data = G.get_edge_data(path[i], path[i + 1])
            interaction_details.append(f"{path[i]} → ({edge_data['effect']}) → {path[i + 1]}")
        return f"Potential interaction path:\n" + "\n".join(interaction_details)
    else:
        return f"No direct interaction found between {drug1} and {drug2}"
```

Visualizing the Knowledge Graph

Visualization is a key aspect of knowledge graph analysis, allowing researchers and healthcare professionals to intuitively understand the complex web of drug interactions. A well-designed visual representation can reveal patterns and relationships that might not be immediately apparent from textual data alone.

```python
import matplotlib.pyplot as plt

def visualize_graph(G):
    pos = nx.spring_layout(G)   # Generate layout for nodes
    plt.figure(figsize=(10, 8))

    # Draw nodes with labels
    nx.draw_networkx_nodes(G, pos, node_color="lightblue", node_size=2000)
    nx.draw_networkx_labels(G, pos, font_size=10, font_color="black")

    # Draw edges with labels
    nx.draw_networkx_edges(G, pos, arrowstyle="->", arrowsize=15)
    edge_labels = nx.get_edge_attributes(G, "effect")
    nx.draw_networkx_edge_labels(G, pos, edge_labels=edge_labels, font_size=8)

    plt.title("Drug-Drug Interaction Knowledge Graph")
    plt.axis("off")
    plt.show()

# Call the visualization function
visualize_graph(G)
```

Explanation:

1. Graph layout:
   - nx.spring_layout: positions nodes to minimize edge overlap
2. Node visualization:
   - draws nodes with a light blue color and labels
3. Edge visualization:
   - draws directed edges with arrowheads
   - labels edges with their effects (e.g., "increased_exposure")
4. Graph title:
   - adds a title to describe the graph

Important Features of the Code

1. Knowledge graph:
   - models drugs, their mechanisms, and interactions as nodes and edges
2. Interaction query:
   - finds paths between drugs and describes interaction effects
3. Visualization:
   - provides an intuitive view of the graph with labeled nodes and edges

Applications

- Pharmacology: analyze drug interactions and their effects
- Clinical decision support: identify potential adverse interactions
- Drug development: explore mechanisms and relationships between compounds

Hybrid AI Approach for Patient Risk Stratification

This example combines machine learning (random forest classifier) with knowledge-based rules to create a hybrid AI approach for patient risk stratification. The machine learning model provides data-driven predictions, while the expert-defined rules adjust these predictions based on domain-specific knowledge, enhancing the interpretability and accuracy of the system.

Data Loading and Preprocessing

The patient data is loaded from .npy files for both training and testing. The features (X_train, X_test) represent patient data, such as vital signs, demographic details, or lab results, while the labels (y_train, y_test) indicate the corresponding risk levels (e.g., "low_risk," "high_risk").

```
import numpy as np
from sklearn.ensemble import RandomForestClassifier
from sklearn.metrics import accuracy_score

# Example data loading (replace with actual file paths)
X_train = np.load("path/to/train/features.npy")
y_train = np.load("path/to/train/labels.npy")
X_test = np.load("path/to/test/features.npy")
y_test = np.load("path/to/test/labels.npy")

# NOTE: Ensure X_train/X_test contain columns for the features used by rules.
# For example, 'history_heart_disease' must be accessible for rule application.

# Example check:
print("Train data shape:", X_train.shape)
print("Test data shape:", X_test.shape)
```

Ensure that the data has been preprocessed (e.g., handling missing values, encoding categorical variables, scaling numerical features) before training the model.

Training a Random Forest Classifier

A random forest classifier, an ensemble learning method, is used for its robustness, ability to handle both categorical and numerical data, and interpretability. The model is trained using the training data, and predictions are made on the test set.

```
# Train a Random Forest model
rf_model = RandomForestClassifier(n_estimators=100, random_state=42)
rf_model.fit(X_train, y_train)

# Predict on test set
y_pred_rf = rf_model.predict(X_test)

# Evaluate baseline accuracy
rf_accuracy = accuracy_score(y_test, y_pred_rf)
print(f"Random Forest accuracy: {rf_accuracy:.3f}")
This provides a baseline machine learning-driven prediction.
```

Incorporating Knowledge-Based Rules

The predictions from the random forest model are refined using expert-defined rules. For instance, a rule might escalate a patient's risk if they have a history of heart disease, even if the model predicts a "high_risk" label. This allows the system to integrate clinical knowledge and improve its decision-making process.

```
def apply_expert_rules(patient_features, predicted_label):
    """
    Example rule-based escalation:
```

```
    - If the patient has a history of heart disease and the model says 'high_risk',
        we upgrade to 'critical_risk'.
    """
    if patient_features["history_heart_disease"] and predicted_label == "high_risk":
        return "critical_risk"
    else:
        return predicted_label


# Apply the hybrid approach
y_pred_hybrid = []
for i in range(len(X_test)):
    # 'X_test[i]' assumed to be a dictionary-like structure or named array
    # with an entry for 'history_heart_disease'.
    # If you're using numpy arrays, map indices or convert to a dict beforehand.
    refined_label = apply_expert_rules(X_test[i], y_pred_rf[i])
    y_pred_hybrid.append(refined_label)


# Evaluate the hybrid model accuracy
hybrid_accuracy = accuracy_score(y_test, y_pred_hybrid)
print(f"Hybrid AI accuracy: {hybrid_accuracy:.3f}")

# Note: If your test data (X_test) is a NumPy array, you'll likely need to convert it into a dictionary-like format or DataFrame to properly reference columns
```

```
like "history_heart_disease". One approach is to
keep a list of feature names (e.g., feature_names =
["history_heart_disease", ...]) and refer to them by
index.
```

In the context of the code, the function apply_expert_rules is applied to each test sample to produce refined predictions.

Evaluating the Hybrid Model

After applying the rules, the refined predictions are compared to the ground truth labels to evaluate the hybrid model's accuracy. This step demonstrates the potential improvement in predictive performance due to the inclusion of expert knowledge.

```
y_pred_hybrid = [apply_expert_rules(X_test[i], y_pred_
rf[i]) for i in range(len(X_test))]
print(f"Hybrid AI accuracy: {accuracy_score(y_test,
y_pred_hybrid):.3f}")
```

Note that the X_test data must be a structured array or dataframe that allows access to specific features like "history_heart_disease" for rule evaluation.

These code snippets provide a starting point for implementing AI techniques in a CDSS. However, it is important to note that these examples are simplified and would need to be adapted and extended based on the specific requirements and data characteristics of a given CDSS application. Additionally, proper data preprocessing, model validation, and testing are crucial steps in developing robust and reliable AI-powered CDSS.

Considerations for the Hybrid AI Approach

When considering the hybrid AI approach, several factors must be addressed to ensure both accuracy and sustainability. First, maintenance and updates play a crucial role. Model updates may require periodic retraining or refinement of the random forest to incorporate new data and adjust to changing patient populations. At the same time, the expert rules embedded in the system should be carefully monitored and revised to reflect the latest clinical knowledge, such as new guidelines or diagnostic criteria.

Another important aspect is the risk of overreliance. In situations where machine learning outputs conflict with rule-based logic, a clear conflict resolution mechanism is needed, whether that entails giving precedence to one component, establishing a consensus rule, or applying more advanced logic. Human oversight remains indispensable, as clinicians must be able to override automated decisions if critical discrepancies are identified or if unique patient circumstances arise.

System complexity also increases as rules accumulate, risking unintended contradictions or overlaps. This phenomenon, sometimes called "rule explosion," can lead to logical conflicts and add to the computational overhead. Especially in real-time clinical environments, processing large sets of rules can slow system performance if not efficiently managed.

Finally, the clinical context must always be at the forefront. Data availability is paramount, as features that inform expert rules—such as "history_heart_disease"—must be routinely and reliably recorded in the EHR. Additionally, compliance with relevant privacy and ethical regulations (for instance, HIPAA in the United States or GDPR in Europe) is essential, as these safeguards ensure that patient data is used responsibly while maintaining the integrity of clinical decision support.

Strengths of the Hybrid Model

The hybrid model brings several key strengths to clinical decision support. First, improved interpretability is achieved through rule transparency and contextual nuances. Each rule is expressed in a human-readable format that encodes clinical knowledge, making it straightforward for clinicians to see the rationale behind particular predictions. Moreover, when the system escalates a case to "critical_risk" due to known cardiac history, healthcare providers can easily trace that decision back to a specific rule, thus enhancing trust and confidence in the system.

A hybrid model also offers enhanced accuracy by leveraging both data-driven insights and domain expertise. Outlier cases or underrepresented scenarios in the training data may be missed by a purely machine-learning-based approach; however, expert rules can fill these gaps and catch critical "red flags." As a result, patients at higher risk are less likely to be overlooked. Additionally, domain experts can precisely specify important conditions or risk adjustments—like certain lab value thresholds—without requiring the model to relearn these insights from scratch.

Finally, the hybrid setup provides considerable flexibility. Since the machine learning component is modular, you can substitute the random forest with another model, such as XGBoost or neural networks, while keeping the same rule-based logic intact. This design also facilitates rule evolution. New guidelines or biomarkers can be incorporated by updating the rules alone, without retraining the entire machine learning pipeline. This adaptability proves especially beneficial in a rapidly changing medical landscape, where clinical best practices and knowledge may shift on relatively short notice.

Clinician Trust

Clinicians may be more inclined to trust hybrid AI systems because they integrate human-like reasoning with explicit rules, such as considering risk level Y for condition X. This alignment with established practice patterns often leads to higher adoption rates, as systems with knowledge-based logic feel more intuitive to healthcare professionals.

These approaches face several challenges. One concern is rule conflicts, wherein the machine learning model might classify a patient as "low_risk" while the rules indicate "high_risk." A clear hierarchy or tie-breaking mechanism must be in place to resolve such contradictions. Another issue is scalability: as clinical knowledge expands, the ruleset can become unwieldy. Maintenance and validation efforts also grow more complex over time. Frequent rule revisions may cause "rule drift," where logic drifts from its original design, and each rule must be tested for potential negative interactions.

Regarding data requirements, a rich feature set is vital. Critical risk factors—such as medical history, vital signs, or lab results—must be thoroughly captured, consistently labeled, and made accessible to both the machine learning model and the rules layer. The system further relies on sufficient labeled data for proper training. Quality labels (e.g., "low_risk" and "high_risk") ensure that the random forest can learn effectively, while rare conditions underscore the importance of expert rules to fill gaps in the training data. Additionally, data integration from various EHR modules and other sources must be managed by a robust pipeline to provide reliable inputs to the hybrid system.

Looking ahead, there are many future enhancements for hybrid models. Explainable AI (XAI) techniques such as SHAP or LIME can highlight which features most influenced the random forest's output, while rule-based insights make the overall decision process more transparent. Automated rule discovery through symbolic AI or pattern mining could also identify novel patterns in patient data, translating them into new rules over time. For adaptive learning, continuous retraining and active learning strategies can improve model performance based on real-world feedback, while large-scale validation—including clinical trials and implementation science—can confirm the model's impact on outcomes like reduced readmissions.

This hybrid AI framework utilizes both machine learning (providing scalable, data-driven insights) and knowledge-based rules (offering interpretability and embedded clinical expertise). By combining these elements, healthcare organizations can develop robust, clinically accepted risk stratification tools that evolve as new data and knowledge become available. Although patient risk stratification is a prominent example, the same concept extends to other clinical decision support tasks, including treatment recommendation, diagnostic assistance, and resource allocation—ensuring clinicians always have a clear, expert-informed AI system at their disposal.

6.4.2 Evaluating CDSS Effectiveness

Evaluating the effectiveness of Clinical Decision Support Systems (CDSSs) is an essential step to ensure that they provide accurate, reliable, and clinically relevant recommendations that improve patient outcomes and healthcare quality. Rigorous evaluation is necessary to justify the investment in CDSS development and implementation, and to identify areas for improvement and refinement.

There are several aspects to consider when evaluating CDSS effectiveness. We discuss several of these below.

Clinical performance metrics: The primary goal of a CDSS is to improve clinical outcomes, such as reducing diagnostic errors, improving treatment selection, and enhancing patient safety. To assess the clinical impact of a CDSS, researchers can measure relevant performance metrics before and after the implementation of the system. These metrics may include the following:

- diagnostic accuracy and specificity
- adherence to clinical guidelines and best practices
- rates of adverse drug events and other patient safety indicators
- patient outcomes, such as mortality, morbidity, and quality of life

User acceptance and satisfaction: The success of a CDSS depends heavily on the willingness of healthcare professionals to use and trust the system. User acceptance and satisfaction can be evaluated through surveys, interviews, and usability testing. Important factors to assess include the following:

- perceived usefulness and ease of use of the CDSS
- integration with clinical workflows and existing health IT systems
- quality and relevance of the CDSS recommendations
- impact on cognitive workload and decision-making efficiency

Economic impact: Implementing a CDSS requires significant financial investments, including costs for development, infrastructure, training, and maintenance. To justify these investments, it is important to evaluate the economic impact of a CDSS, such as

- cost savings from reduced diagnostic testing, hospital readmissions, and adverse events
- improvements in operational efficiency and resource utilization
- return on investment (ROI) and cost-effectiveness analyses

While CDSSs are designed to improve clinical decision-making, they may also introduce unintended consequences that need to be carefully monitored and addressed. These may include the following:

- alert fatigue and desensitization to CDSS recommendations
- overreliance on technology and reduced critical thinking skills
- perpetuation of biases or errors in the underlying clinical knowledge base
- widening of health disparities due to differential access or use of a CDSS

To conduct a comprehensive evaluation of a CDSS's effectiveness, a combination of quantitative and qualitative methods is often necessary. This may involve the following:

- randomized controlled trials (RCTs) to compare clinical outcomes between a CDSS-supported and usual care groups
- observational studies to assess the real-world impact of a CDSS on clinical practice and patient outcomes

- qualitative studies, such as focus groups and interviews, to gather user feedback and identify barriers and facilitators to CDSS adoption
- economic evaluations, such as cost-benefit and cost-effectiveness analyses, to assess the financial impact of CDSS
- simulation studies to test the performance of a CDSS under different clinical scenarios and edge cases

Practical Tip: When planning an evaluation of a CDSS's effectiveness, it is important to engage a multidisciplinary team of stakeholders, including clinicians, informaticians, data scientists, and health services researchers. This team should collaborate to define clear evaluation objectives, select appropriate methods and metrics, and interpret the results in the context of the specific healthcare setting and patient population.

It is also important to recognize that the evaluation of a CDSS's effectiveness is an ongoing process, rather than a one-time event. As clinical knowledge improves and new data becomes available, the CDSS needs to be continuously updated and re-evaluated to ensure its continued relevance and effectiveness.

Evaluating the effectiveness of a CDSS is a complex but essential task that requires a comprehensive and multidisciplinary approach. By rigorously assessing the clinical, user, economic, and unintended impacts of a CDSS, healthcare organizations can make informed decisions about the development, implementation, and refinement of these powerful tools to support clinical decision-making and improve patient care.

6.5 DATA GOVERNANCE AND REGULATORY COMPLIANCE

As healthcare organizations increasingly rely on clinical decision support systems (CDSSs) and other AI-powered tools, ensuring the proper governance and regulatory compliance of patient data becomes paramount. *Data governance* refers to the overall management of the availability, usability, integrity, and security of the data employed in an organization. In the context of healthcare, this involves establishing policies, procedures, and standards to ensure that patient data is collected, stored, and used in a manner that is consistent with ethical principles, legal requirements, and best practices for data management.

Regulatory compliance is a critical component of data governance in healthcare. In the United States, the Health Insurance Portability and Accountability Act (HIPAA) sets national standards for the protection of sensitive patient health information. HIPAA's Privacy Rule regulates the use and disclosure of protected health information (PHI), while the Security Rule establishes national standards for the security of electronic PHI. Healthcare organizations must implement appropriate administrative, physical, and technical safeguards to ensure the confidentiality, integrity, and availability of PHI.

In the European Union, the General Data Protection Regulation (GDPR) sets strict requirements for the collection, storage, and use of personal data, including health data. Under GDPR, healthcare organizations must obtain explicit consent from patients for the processing of their data, provide patients with access to their data, and ensure that patient data is used only for the specific purposes for which it was collected.

To meet these regulatory requirements and establish effective data governance practices, healthcare organizations should consider the following critical elements:

1. Data inventory and mapping: Organizations should conduct a comprehensive inventory of all patient data assets, including structured and unstructured data, and map the flow of data across different systems and stakeholders. This helps identify potential risks and vulnerabilities in the data life cycle.

2. Data classification and risk assessment: Patient data should be classified based on its sensitivity and criticality, and potential risks associated with each data class should be assessed. This helps prioritize data protection efforts and allocate resources appropriately.

3. Data access control and authentication: Strict access controls should be implemented to ensure that only authorized personnel can access patient data on a need-to-know basis. This may involve the use of role-based access control (RBAC), multi-factor authentication (MFA), and other security measures.

4. Data encryption and secure storage: Patient data should be encrypted both at rest and in transit to protect against unauthorized access or breaches. Secure storage solutions, such as encrypted databases and cloud storage with strong access controls, should be used to safeguard data.

5. Data backup and disaster recovery: Regular data backups should be performed, and disaster recovery plans should be established to ensure the availability and integrity of patient data in the event of a system failure or security incident.

6. Data retention and disposal: Organizations should establish policies for the retention and disposal of patient data in accordance with legal requirements and best practices. This may involve secure data destruction methods and documentation of data disposal activities.

7. Data use agreements and consent management: When sharing patient data with external parties, such as research institutions or third-party vendors, organizations should establish clear data use agreements that specify the terms and conditions of data access and use. Patients' consent for data sharing should be obtained and managed in accordance with applicable regulations.

8. Training and awareness: All personnel involved in the handling of patient data should receive regular training on data governance policies, procedures, and best practices. This helps ensure that everyone understands their roles and responsibilities in protecting patient data.

Practical Tip: Conducting regular audits and assessments of data governance practices can help identify gaps and areas for improvement. This may involve the use of automated tools for data discovery and classification, as well as manual reviews of policies and procedures. Engaging with external experts, such as legal counsel and cybersecurity consultants, can provide valuable guidance and support in ensuring regulatory compliance and implementing best practices for data governance.

By establishing robust data governance practices and ensuring regulatory compliance, healthcare organizations can build trust with patients, mitigate risks associated with data breaches and misuse, and create a solid foundation for the ethical and responsible use of AI and CDSSs in patient care.

6.5.1 AI in Data Protection and Anonymization

AI is playing an increasingly important role in enhancing data protection and privacy in healthcare. As clinical decision support systems (CDSSs) and other AI-powered tools rely on vast amounts of patient data, ensuring the security and confidentiality of this sensitive information is crucial. AI techniques can be utilized to safeguard patient data through advanced anonymization methods, intrusion detection, and secure data sharing.

Data anonymization is the process of removing personally identifiable information (PII) from datasets, such that individual patients cannot be identified. This is essential for protecting patient privacy when sharing data with third parties, such as researchers or AI developers. Traditional anonymization techniques, such as data masking or tokenization, can be time-consuming and may not always provide sufficient protection against re-identification attacks.

AI-powered anonymization techniques offer a more robust and scalable solution. For example, generative adversarial networks (GANs) can be used to create synthetic datasets that mimic the statistical properties of the original data without revealing any real patient information. These synthetic datasets can be used for training AI models or conducting research without compromising patient privacy.

Another promising approach is the use of differential privacy, a mathematical framework that allows for the controlled release of aggregate statistics while minimizing the risk of individual identification. Differential privacy techniques, such as the addition of carefully calibrated noise to query results, can be combined with AI algorithms to enable privacy-preserving data analysis and machine learning.

AI can also be used to detect and prevent data breaches or unauthorized access to patient data. Machine learning algorithms can be trained to identify anomalous patterns of data access or network traffic that may indicate a potential security threat. For example, unsupervised learning techniques, such as clustering or anomaly detection, can flag unusual user behavior or system events for further investigation.

In addition, AI can enable secure data sharing and collaboration among healthcare organizations and research institutions. Federated learning is an approach that allows multiple parties to train machine learning models on their local data without sharing the raw data itself. Each party trains a local model on their own data and shares only the model updates with a central server, which aggregates the updates to create a global model. This allows for the benefits of collaborative learning while preserving data privacy and security.

Practical considerations for implementing AI in data protection and anonymization include the following:

1. Data quality and preprocessing: Ensuring that patient data is accurate, complete, and properly formatted is essential for the success of AI-powered data protection techniques. This may involve data cleaning, harmonization, and standardization processes to prepare the data for analysis.

2. Model interpretability and transparency: When using AI for data protection, it is important to ensure that the models are interpretable and transparent. This allows for the identification of potential biases or errors in the anonymization process and helps build trust among stakeholders.

3. Regulatory compliance: The use of AI in data protection must comply with relevant regulations, such as HIPAA and GDPR. This may involve conducting privacy impact assessments, obtaining patient consent for data processing, and implementing appropriate technical and organizational measures to ensure data security.

4. Continuous monitoring and improvement: As data protection threats evolve over time, it is important to continuously monitor the performance of AI-powered data protection systems and update them as needed. This may involve regular security audits, penetration testing, and the incorporation of new AI techniques as they become available.

Practical Tip: When implementing AI for data protection and anonymization, it is important to engage with a multidisciplinary team of experts, including data scientists, cybersecurity professionals, legal experts, and healthcare domain experts. This helps ensure that the AI solutions are technically sound, legally compliant, and aligned with the specific needs and constraints of the healthcare organization.

By leveraging AI techniques for data protection and anonymization, healthcare organizations can enhance the security and privacy of patient data while enabling the responsible use of this data for research, innovation, and the development of more effective CDSS. As AI in healthcare changes, it will be essential to balance the benefits of data-driven insights and the imperative to protect patient privacy and maintain trust in the healthcare system.

Code Tip

Here is a detailed code snippet that demonstrates the use of AI techniques for data protection and anonymization in healthcare.

```
import numpy as np
import pandas as pd
from sklearn.datasets import load_iris
from sklearn.model_selection import train_test_split
from sklearn.preprocessing import StandardScaler
from sklearn.linear_model import LogisticRegression
from sklearn.metrics import accuracy_score
from opendp.smartnoise.synthesizers.pytorch.pytorch_synthesizer import PytorchDPSynthesizer
from opendp.smartnoise.synthesizers.preprocessors.preprocessing import GeneralTransformer


# Load and preprocess the iris dataset (as a proxy for healthcare data)
iris = load_iris()
X, y = iris.data, iris.target
X_train, X_test, y_train, y_test = train_test_split(X, y, test_size=0.2, random_state=42)


# Scale the features
scaler = StandardScaler()
X_train_scaled = scaler.fit_transform(X_train)
X_test_scaled = scaler.transform(X_test)
```

```python
# Train a logistic regression model on the original data
model = LogisticRegression(random_state=42)
model.fit(X_train_scaled, y_train)
y_pred = model.predict(X_test_scaled)
print(f"Accuracy on original data: {accuracy_score(y_test, y_pred):.2f}")


# Generate synthetic data using differential privacy
preprocessor = GeneralTransformer(epsilon=1.0)
preprocessor.fit(X_train)
X_train_preprocessed = preprocessor.transform(X_train)

synthesizer = PytorchDPSynthesizer(preprocessor=preprocessor)
synthesizer.fit(X_train_preprocessed, y_train)
X_synthetic, y_synthetic = synthesizer.sample(n_samples=len(X_train))


# Train a logistic regression model on the synthetic data
model_synthetic = LogisticRegression(random_state=42)
model_synthetic.fit(X_synthetic, y_synthetic)
y_pred_synthetic = model_synthetic.predict(X_test_scaled)
print(f"Accuracy on synthetic data: {accuracy_score(y_test, y_pred_synthetic):.2f}")
```

```python
# Simulate a data breach and detect anomalies using AI
def simulate_data_breach(X):
    # Randomly select a subset of the data to be "breached"
    breach_indices = np.random.choice(len(X), size=int(0.1 * len(X)), replace=False)
    X_breached = X.copy()
    # Add random noise to the breached data to simulate anomalous behavior
    X_breached[breach_indices] += np.random.normal(loc=0, scale=1, size=(len(breach_indices), X.shape[1]))
    return X_breached


X_breached = simulate_data_breach(X_test_scaled)


# Use an unsupervised anomaly detection algorithm (e.g., Isolation Forest) to detect breaches
from sklearn.ensemble import IsolationForest

detector = IsolationForest(contamination=0.1, random_state=42)
detector.fit(X_test_scaled)
anomalies = detector.predict(X_breached)
print(f"Number of anomalies detected: {len(anomalies[anomalies == -1])}")
```

This code snippet demonstrates several key concepts related to AI in data protection and anonymization:

- Data preprocessing: The code loads the iris dataset (as a proxy for healthcare data) and performs standard preprocessing steps, such as splitting the data into train and test sets and scaling the features.

- Training a model on the original data: A logistic regression model is trained on the original data to establish a baseline accuracy.

- Generating synthetic data using differential privacy: The code uses the OpenDP library to generate synthetic data that mimics the statistical properties of the original data while preserving privacy. The GeneralTransformer is used to preprocess the data, and the PytorchDPSynthesizer is used to generate the synthetic data.

- Training a model on the synthetic data: A logistic regression model is trained on the synthetic data and evaluated on the test set to assess the utility of the synthetic data.

- Simulating a data breach and detecting anomalies: The code simulates a data breach by randomly selecting a subset of the test data and adding random noise to it. An unsupervised anomaly detection algorithm (Isolation Forest) is then used to detect the anomalous behavior in the breached data.

This code snippet provides a starting point for implementing AI techniques for data protection and anonymization in healthcare. However, it is important to note that this is a simplified example, and real-world applications would require more extensive data preprocessing, model selection, and evaluation, as well as compliance with relevant regulations and best practices for data governance and security.

CHAPTER 7

From Bench to Bedside: Translating AI into Practice

7.1 FROM BENCH TO BEDSIDE: TRANSLATING AI INTO PRACTICE

The journey from developing an AI model in a research lab to deploying it in a real-world clinical setting is complex and multifaceted. It requires a deep understanding of the clinical context, intended use case, available data and infrastructure, and potential risks and benefits. Translating AI into practice involves navigating a series of technical, operational, and ethical challenges, each of which requires careful consideration and collaboration among diverse stakeholders.

One of the first steps in this process is defining the clinical problem that the AI model is intended to address. This requires close collaboration between AI researchers and healthcare providers to ensure that the model is grounded in a deep understanding of the clinical workflow, the patient population, and the desired outcomes. It also involves identifying the important data sources and variables that will be used to train and validate the model, as well as the performance metrics that will be used to assess its effectiveness.

Once the clinical problem and data sources have been defined, the next step is to develop and train the AI model using appropriate machine learning techniques. This often involves a process of iterative refinement and validation, where the model is tested on different subsets of the data to assess its performance and identify areas for improvement. It may also involve techniques such as data augmentation, transfer learning, or ensemble modeling to improve the model's robustness and generalizability.

However, developing a high-performing AI model is only the first step in the translation process. To be successfully deployed in a clinical setting, the model must also be integrated into existing workflows and systems in a way that is seamless, efficient, and user-friendly. This requires close collaboration between AI developers, IT staff, and clinical end-users to ensure that the model is properly validated, monitored, and maintained over time.

One challenge in this process is ensuring that the AI model is able to handle the complexity and variability of real-world clinical data. Unlike the carefully curated datasets used in research settings, clinical data is often messy, incomplete, and inconsistent, with many potential sources of bias and error. To address this challenge, AI models must be designed with robust data preprocessing and quality control mechanisms, as well as the ability to adapt to changing data inputs and clinical contexts.

Another challenge is ensuring that the AI model is able to communicate its results and recommendations in a way that is meaningful and actionable for clinical end-users. This may involve developing intuitive user interfaces, providing clear explanations of the model's reasoning and limitations, and integrating the model's outputs with existing clinical decision support systems and workflows.

Beyond these technical challenges, there are also important ethical and regulatory considerations that must be addressed in the translation of AI into clinical practice. These include issues of data privacy and security, informed consent, algorithmic bias and fairness, and liability and accountability for AI-based decisions. Addressing these issues requires ongoing collaboration and dialogue among healthcare providers, technology developers, policymakers, and patients to ensure that AI is developed and deployed in a way that is transparent, accountable, and aligned with societal values and priorities.

Despite these challenges, the potential benefits of AI in healthcare are significant and far-reaching. By enabling more precise, personalized, and proactive care, AI has the potential to improve patient outcomes, reduce healthcare costs, and enhance the efficiency and effectiveness of clinical workflows. As AI is utilized more often in healthcare, it is essential to approach the translation process with collaboration, innovation, and ethical responsibility, working together to ensure that the benefits of this transformative technology are realized for patients and society as a whole.

Here is a basic Python code snippet that demonstrates the process of loading a trained AI model and using it to make predictions on new clinical data. We build upon this example throughout the chapter to illustrate various aspects of AI integration into clinical workflows.

```
import numpy as np
import pandas as pd
from sklearn.externals import joblib


# Load the trained AI model
model = joblib.load('path/to/trained/model.pkl')


# Load and preprocess the new clinical data
data = pd.read_csv('path/to/clinical/data.csv')


data_preprocessed = preprocess_data(data)


# Custom preprocessing function


# Make predictions using the loaded model
predictions = model.predict(data_preprocessed)


# Postprocess the predictions and integrate with
clinical workflow
postprocessed_predictions =
postprocess_predictions(predictions)
# Custom postprocessing function
integrate_with_clinical_workflow(postprocessed_
predictions)   # Custom integration function
```

In this example, we first load a trained AI model using the joblib library, which is a common tool for serializing and deserializing Python objects, including trained machine learning models.

Next, we load the new clinical data from a CSV file and preprocess it using a custom preprocess_data function. This function would typically involve tasks such as data cleaning, feature scaling, and handling missing values, depending on the specific requirements of the AI model.

We then use the loaded AI model to make predictions on the preprocessed clinical data using the predict method.

Finally, we postprocess the raw predictions using a custom postprocess_predictions function, which may involve tasks such as thresholding, calibration, or formatting the predictions for integration with clinical systems. The postprocessed predictions are then passed to a custom integrate_with_clinical_workflow function, which would handle the actual integration of the AI model outputs with the existing clinical workflows and systems.

Throughout the rest of this chapter, we expand upon this basic example to demonstrate various techniques and considerations for data preprocessing, model validation, user interface design, and ethical safeguards in the context of AI integration into clinical practice.

7.2   IMAGING ANALYSIS AND INTERPRETATION

Medical imaging is one of the most promising areas for AI application in healthcare. Deep learning algorithms, particularly convolutional neural networks (CNNs), have demonstrated remarkable performance in analyzing and interpreting various types of medical images, including radiographs, CT scans, MRIs, and pathology slides. By learning to recognize complex patterns and features in these images, AI models can assist clinicians in detecting abnormalities, classifying diseases, and guiding treatment decisions.

In radiology, AI models can be trained to detect and characterize a wide range of abnormalities, such as lung nodules, brain tumors, and cardiovascular diseases. For example, a CNN model trained on a large dataset of chest X-rays can learn to identify signs of pneumonia, tuberculosis, or lung cancer, and highlight these regions for further review by radiologists. Similarly, AI models can be used to segment and quantify brain lesions on MRI scans, helping to monitor disease progression and treatment response in conditions such as multiple sclerosis and Alzheimer's disease.

AI is also transforming the field of digital pathology, where high-resolution images of tissue samples are analyzed to diagnose and classify diseases such as cancer. By learning to recognize morphological and molecular features that are predictive of cancer subtypes and prognosis, AI models can assist pathologists in making more accurate and efficient diagnoses, and guide personalized treatment decisions.

Practical Tip: When developing AI models for imaging analysis, it is important to use large, diverse, and well-curated datasets for training and validation. This helps to ensure that the models are robust to variations in image quality, acquisition protocols, and patient populations. It is also important to involve expert clinicians in the annotation and interpretation of the training data to ensure that the models are learning clinically relevant features and not just artifacts or noise.

Here is an example of how to use a pretrained CNN model to classify chest X-ray images as normal or pneumonia:

1. Install TorchXRayVision.

```
pip install torchxrayvision
```

NOTE *Make sure you have a compatible version of PyTorch installed. If you have not installed PyTorch yet, visit pytorch.org for instructions.*

2. Code Example

```
import torch
import torchxrayvision as xrv
from PIL import Image
import torchvision.transforms as transforms

# 1. Load the pre-trained DenseNet model (trained on NIH ChestX-ray14 dataset)

model = xrv.models.DenseNet(weights="nih")
model.eval()   # set model to inference mode
```

```python
# 2. Path to your chest X-ray
img_path = 'path/to/chest_xray.jpg'

# 3. Load and preprocess the image
pil_image = Image.open(img_path).convert("RGB")
transform = transforms.Compose([
    transforms.Resize((224, 224)),
        transforms.ToTensor(),
        transforms.Normalize(mean=[0.485, 0.456, 0.406],
                    std=[0.229, 0.224, 0.225])
        ])
input_tensor = transform(pil_image).unsqueeze(0)  # shape: (1, 3, 224, 224)

# 4. Predict
with torch.no_grad():
        preds = model(input_tensor)  # shape: (1, number_of_pathologies)

# 5. Find the probability for pneumonia

# TorchXRayVision models output unnormalized logits, so apply sigmoid
pneumonia_idx = model.pathologies.index("Pneumonia")  # returns an index
pneumonia_logit = preds[0, pneumonia_idx]
pneumonia_prob = torch.sigmoid(pneumonia_logit).item()
```

```
# 6. Threshold and print result
threshold = 0.5


if pneumonia_prob > threshold:
    print(f"Pneumonia detected! Probability: {pneumonia_prob:.2f}")
else:
print(f"Normal chest X-ray (or pneumonia less likely). Probability: {pneumonia_prob:.2f}")
```

How it works

- xrv.models.DenseNet(weights="nih") loads a DenseNet model pre-trained on the NIH ChestX-ray14 dataset, which includes a label for "Pneumonia" among other common chest pathologies.
- We resize the image to (224, 224) and apply standard PyTorch ImageNet normalization (mean [0.485,0.456,0.406] and std [0.229,0.224,0.225]).
- We then apply a sigmoid function to the raw model output for pneumonia because the model typically returns logits (unnormalized scores).
- Finally, we check if the pneumonia probability is above a threshold (commonly 0.5) to decide if pneumonia is detected.

Additional Notes:

1. Model pathologies
   - To see all pathologies that TorchXRayVision is predicting, you can print model.pathologies.
2. Adjusting the threshold
   - A threshold of 0.5 is arbitrary; it may not be optimal in every setting.
   - For research/clinical usage, you should determine an appropriate cut-off based on the validation data and desired sensitivity/specificity.

3. Fine-tuning

    - If you have a specific dataset with labels that differ from the NIH ChestX-ray14 labeling scheme, consider fine-tuning the model on your own data.

4. Clinical deployment

    - Remember that these models are research tools and not intended as a replacement for clinician diagnosis. If you plan to use them in a real clinical workflow, they must be properly validated and cleared according to the applicable policies and regulations.

You now have code that loads a publicly available chest X-ray CNN (via TorchXRayVision), preprocesses a new X-ray image, and makes a pneumonia detection prediction.

7.2.1 Pathology and Lab Result Processing

In addition to imaging, AI can also be applied to the analysis and interpretation of other types of diagnostic data, such as pathology reports and laboratory test results. These data sources often contain rich information about a patient's health status and disease progression, but can be challenging to analyze and interpret due to their complexity, variability, and volume.

Pathology reports, for example, are often written in free-text format and may contain a wide range of morphological and molecular features that are relevant to cancer diagnosis and prognosis. By using natural language processing (NLP) techniques, such as named entity recognition and sentiment analysis, AI models can extract structured information from these reports and identify key features that are predictive of patient outcomes. This can help to automate and standardize the coding and classification of pathology diagnoses, reducing the workload of pathologists and improving the accuracy and consistency of reporting.

Similarly, laboratory test results, such as blood counts, chemistry panels, and biomarker assays, can provide valuable information about a patient's health status and disease risk. However, interpreting these results often requires considering multiple parameters and their interactions, as well as the patient's clinical history and demographics. By using machine learning algorithms, such as decision trees, random forests, and support vector machines, AI models can learn to identify patterns and relationships in laboratory data that are predictive of disease onset, progression, and treatment response. This can help to alert clinicians to potential problems, such as acute kidney injury or sepsis, and guide personalized treatment decisions.

Practical Tip: When developing AI models for pathology and laboratory data, it is important to work closely with domain experts, such as pathologists and laboratory scientists, to ensure that the models are clinically relevant and interpretable. This may involve incorporating domain-specific knowledge and constraints into the model design, such as the use of standardized terminologies and ontologies, or the inclusion of rule-based logic to capture expert reasoning. It is also important to validate the models on independent datasets and to assess their performance in the context of real-world clinical workflows.

```
pip install spacy scispacy
pip install https://s3-us-west-2.amazonaws.com/ai2-s2-scispacy/releases/en_core_sci_md-0.5.1.tar.gz

Note: The above URL corresponds to a specific version of en_core_sci_md. You can check the scispaCy GitHub for updated models.
import spacy
from spacy.pipeline import EntityRuler


# 1. Load a publicly available scientific/biomedical model
nlp = spacy.load('en_core_sci_md')


# 2. Create an EntityRuler to identify domain-specific patterns
ruler = EntityRuler(nlp, overwrite_ents=True)


# 3. Define patterns for domain-specific labels (DIAGNOSIS, SIZE, MARGIN, BIOMARKER)
#    These are simple token-based patterns. You can refine them for more complex scenarios.
```

```python
patterns = [
    {
        "label": "DIAGNOSIS",
        "pattern": [{"LOWER": "invasive"}, {"LOWER": "ductal"}, {"LOWER": "carcinoma"}]
    },
    {
        "label": "SIZE",
        "pattern": [
            {"IS_DIGIT": True, "OP": "+"},    # e.g., 1, 1.5, etc.
            {"LOWER": "cm"}
        ]
    },
    {
        "label": "MARGIN",
        "pattern": [{"LOWER": "margins"}, {"IS_PUNCT": True, "OP": "?"}, {"LOWER": "negative"}]
    },
    {
        "label": "BIOMARKER",
        "pattern": [{"LOWER": {"IN": ["er", "pr", "her2"]}}]
    }
]

ruler.add_patterns(patterns)
```

```python
nlp.add_pipe(ruler, before="ner")  # Insert the custom rules before spaCy's built-in NER


# 4. Define the pathology report text
report_text = """
Diagnosis: Invasive ductal carcinoma, grade 2, measuring 1.5 cm in greatest dimension.
Margins: Negative for tumor.
ER: Positive (90% of cells).
PR: Positive (80% of cells).
HER2: Negative (1+ by IHC).
"""


# 5. Process the text
doc = nlp(report_text)


# 6. Initialize placeholders to store extracted data
diagnosis = None
size = None
margin = None
biomarkers = {}


# 7. Extract relevant entities
for ent in doc.ents:
    if ent.label_ == 'DIAGNOSIS':
        diagnosis = ent.text
    elif ent.label_ == 'SIZE':
```

```
            size = ent.text
    elif ent.label_ == 'MARGIN':
            margin = ent.text
    elif ent.label_ == 'BIOMARKER':
        # Attempt to parse the next tokens for positivity/negativity and any percentages
        # This logic is naive; real-world usage often requires more robust patterns
        biomarker_key = ent.text.upper()   # e.g., 'ER', 'PR', 'HER2'
        # Try to look ahead in the doc for the next few tokens
        start_index = ent.end
        next_tokens = doc[start_index:start_index+5]
        biomarkers[biomarker_key] = next_tokens.text.strip()


# 8. Print the extracted information
print(f"Diagnosis: {diagnosis}")
print(f"Size: {size}")
print(f"Margin: {margin}")
for marker, info in biomarkers.items():
    print(f"{marker}: {info}")
```

How This Approach Works

1. scispaCy model
   - We load en_core_sci_md, a publicly available spaCy model specialized for scientific text. This provides better tokenization and vector representations for biomedical content compared to en_core_web_sm.

2. EntityRuler
   - We create an EntityRuler pipeline component that allows us to add rule-based entity patterns for clinical terms (e.g., "invasive ductal carcinoma" as DIAGNOSIS and "ER"/"PR"/"HER2" as BIOMARKER).
   - This is often easier to maintain and more transparent than training a fully custom NER model from scratch, especially if you have stable terminology patterns you want to capture.

3. Simple patterns
   - For example, {"LOWER": "invasive"}, {"LOWER": "ductal"}, {"LOWER": "carcinoma"} looks for the sequence "invasive ductal carcinoma" in lowercase.
   - You can extend or fine-tune these patterns to match your data more accurately (e.g., matching synonyms or variations).

4. Extracting values
   - After the text is processed, we iterate through doc.ents to see which label was assigned (DIAGNOSIS, SIZE, MARGIN, and BIOMARKER).
   - For biomarkers, we do a naive "look ahead" approach to capture positivity or negativity. You might adapt this to parse the exact numeric percentage or more complex statements.

5. Customization
   - Real pathology reports can be more complex than this example. In practice, you would add more patterns and possibly combine rule-based and statistical NER approaches.
   - You might also feed the text through additional components (e.g., regex-based extraction of numeric values, custom classification logic, or a knowledge-based approach for synonyms and abbreviations).

Additional Tips

- Fine-tuning: If your pathology reports have a particular style or set of terminologies, consider creating a comprehensive set of rules or training a custom NER model on annotated data.
- Complex outputs: For advanced tasks like relationship extraction (e.g., linking biomarkers to the exact measurement or positivity status), you may

- need more sophisticated approaches (dependency parsing, transformer-based relation extraction, etc.).
- Model versions: Periodically check for updated scispaCy models or spaCy versions to ensure maximum compatibility and performance.

This is just a simple example, but it illustrates the basic workflow of using NLP techniques to extract structured information from unstructured clinical text. In real-world applications, there would be additional steps for data cleaning, model training and validation, and integration with clinical workflows, as well as considerations for data privacy, model interpretability, and ethical safeguards.

7.3 RARE DISEASE IDENTIFICATION

Rare diseases, defined as conditions affecting fewer than 200,000 individuals in the United States, pose significant challenges for diagnosis and treatment. Patients with rare diseases often face lengthy diagnostic odysseys, misdiagnoses, and limited treatment options, due to the lack of clinical expertise and research on these conditions. AI has the potential to accelerate the diagnosis and management of rare diseases by leveraging large-scale clinical and genomic data to identify patterns and similarities across patients.

One promising approach is to use machine learning algorithms to mine electronic health records (EHRs) and other clinical data sources for patients with similar symptom profiles and clinical histories. By identifying clusters of patients with similar presentations, even if they have not yet been diagnosed with a specific rare disease, AI models can help to surface potential cases and guide targeted genetic testing and other diagnostic evaluations. This can be particularly valuable for rare diseases with nonspecific or heterogeneous presentations, which may be missed by traditional diagnostic algorithms.

Another approach is to use NLP and text mining techniques to extract relevant information from unstructured data sources, such as clinical notes, research publications, and patient registries. By identifying key terms and concepts related to rare diseases, such as specific symptoms, biomarkers, or genetic variants, AI models can help to surface potential cases and guide further investigation. This can be particularly valuable for rare diseases with limited published literature or clinical expertise, where traditional knowledge bases may be incomplete or outdated.

In addition to aiding diagnosis, AI can also be used to guide the management and treatment of rare diseases. By analyzing large-scale clinical and genomic data from patients with similar conditions, AI models can help to identify potential drug targets, predict treatment responses, and optimize care pathways. This can be particularly valuable for rare diseases with limited treatment options or clinical trial data, where personalized approaches may be necessary.

PRACTICAL TIP *When developing AI models for rare disease identification, it is important to collaborate with patient advocacy groups and rare disease experts to ensure that the models are clinically relevant and patient-centered. This may involve incorporating patient-reported outcomes and experiences into the model design, as well as considering the ethical and social implications of rare disease diagnosis and treatment. It is also important to validate the models on diverse and representative patient populations, and to assess their performance in the context of real-world clinical workflows.*

You can use publicly available EHR data. Here is an example of how to use clustering algorithms to identify potential rare disease cases from EHR data.

Please note that the snippet provided is illustrative. In practice, you would need to obtain and preprocess the MIMIC-IV data (or another public dataset such as one from Kaggle) according to your specific project and institutional review board (IRB) requirements.

1. Setup and data acquisition

    a. Download MIMIC-IV from PhysioNet (requires credentialing and a data use agreement).

    b. Convert the relevant tables (e.g., admissions, diagnoses, lab measurements) into a CSV file or a relational database.

    c. Load the CSV into a Pandas DataFrame in Python.

2. Example code using DBSCAN for rare disease detection

```python
import pandas as pd
from sklearn.cluster import DBSCAN
from sklearn.preprocessing import StandardScaler
```

```python
# 1. Load EHR data (publicly available example:
MIMIC-IV CSV, after you extract relevant columns)
ehr_data = pd.read_csv('path/to/mimic_subset.csv')


# 2. Preprocess the data
#    - This might include cleaning, imputation,
encoding categorical variables, etc.
def preprocess_data(df):
    # Example transformations:
    #   - Drop rows with too many missing values
    #   - Fill numeric NaNs with mean/median
    #   - Encode categorical data
    #   - Return the cleaned DataFrame
    df = df.dropna(subset=['symptom1', 'symptom2',
'lab_value1', 'lab_value2'])
    df['symptom1'] = df['symptom1'].astype(float)
    df['symptom2'] = df['symptom2'].astype(float)
    df['lab_value1'] = df['lab_value1'].astype(float)
    df['lab_value2'] = df['lab_value2'].astype(float)
    return df


ehr_data = preprocess_data(ehr_data)


# 3. Select features for clustering (symptoms, lab
values, etc.)
features = ['symptom1', 'symptom2', 'lab_value1',
'lab_value2']
X = ehr_data[features]
```

```python
# 4. Scale the features (DBSCAN is distance-based; scaling often helps)
scaler = StandardScaler()
X_scaled = scaler.fit_transform(X)


# 5. Perform DBSCAN clustering
#    eps and min_samples are hyperparameters; tune them based on your data
dbscan = DBSCAN(eps=0.8, min_samples=5)
clusters = dbscan.fit_predict(X_scaled)


# 6. Identify outliers (label = -1 in DBSCAN) or small clusters
#    DBSCAN assigns cluster labels (0,1,2,...) and -1 to points it considers outliers
ehr_data['cluster_label'] = clusters


# Let's define "rare disease cluster" as either outliers or clusters with small membership
cluster_counts = ehr_data['cluster_label'].value_counts()
rare_labels = cluster_counts[cluster_counts < 10].index  # e.g., cluster size < 10
rare_disease_candidates = ehr_data[ehr_data['cluster_label'].isin(rare_labels)]


# 7. Output these potential rare disease cases for further investigation
rare_disease_candidates.to_csv('path/to/potential_rare_cases.csv', index=False)
```

```
print("Potential rare disease clusters identified:")
print(rare_labels)
print("Number of potential cases:", 
len(rare_disease_candidates))
```

How This Approach Works

1. Publicly available dataset
   - We reference MIMIC-IV from PhysioNet as an example of a large EHR dataset that can be used for research purposes after appropriate credentialing and approvals.

2. DBSCAN
   - DBSCAN stands for Density-Based Spatial Clustering of Applications with Noise. It groups points that are closely packed together, marking points that lie alone in low-density regions as outliers (label = -1).
   - This is particularly useful when you are trying to find small numbers of unusual patients who could represent rare diseases.

3. Preprocessing
   - We show a simplified preprocess_data function. In reality, you would handle missing data, transform/encode categorical variables, and normalize or standardize numeric columns.
   - Feature selection (e.g., which labs or symptoms to include) is crucial and often requires domain expertise.

4. Thresholding "rare" clusters
   - After clustering, we count how many patients are in each cluster. Clusters below a certain threshold (like <10 patients) or the DBSCAN outlier label (-1) might be flagged for deeper manual review.
   - Adjust these thresholds or outlier definitions to match your clinical or research objectives.

5. Next steps
   - Further validate the flagged cases: examine their charts, diagnoses, and outcomes to see if they truly represent rare diseases or if they are just noisy data points.

- You might combine this with an anomaly detection method (e.g., IsolationForest or LocalOutlierFactor) to cross-check results.

By combining a publicly available EHR dataset (e.g., MIMIC-IV) with a robust density-based clustering algorithm like DBSCAN, you can more effectively isolate small patient groups or outliers that may correspond to unusual or rare disease cases.

This is just a simple example, but it illustrates the basic workflow of using clustering algorithms to identify potential rare disease cases from large-scale clinical data. In real-world applications, there would be additional steps for data preprocessing, feature selection, model validation, and integration with clinical workflows, as well as considerations for data privacy, model interpretability, and ethical safeguards.

7.3.1 Personalized Treatment Recommendations

One of the promises of AI in healthcare is the ability to provide personalized treatment recommendations based on a patient's individual characteristics, preferences, and goals. By analyzing large-scale clinical and genomic data, AI models can identify patterns and relationships that are predictive of treatment response and outcomes, and guide the selection of targeted therapies that are tailored to each patient's unique needs.

An enabler of personalized treatment recommendations is the integration of multimodal data sources, such as electronic health records, genomic profiles, imaging data, and patient-reported outcomes. By combining these diverse data types, AI models can develop a more comprehensive and nuanced understanding of each patient's health status, disease trajectory, and treatment options. For example, by analyzing a patient's genomic profile in conjunction with their clinical history and imaging data, AI models can identify specific genetic variants or biomarkers that are predictive of response to targeted therapies, such as kinase inhibitors or immunotherapies.

Another important aspect of personalized treatment recommendations is the incorporation of patient preferences and values into the decision-making process. AI models can be designed to elicit and integrate patient-reported outcomes, such as quality of life measures or treatment goals, into the recommendation algorithms. This can help to ensure that treatment decisions are aligned with each patient's individual priorities and values, and that potential trade-offs between efficacy and side effects are carefully considered.

In addition to guiding initial treatment selection, AI models can also be used to monitor and adapt treatment plans over time based on patient response and changing circumstances. By continuously analyzing patient data and outcomes, AI models can identify early signs of treatment failure or adverse events, and suggest alternative therapies or dosing regimens that may be more effective or tolerable. This can help to optimize treatment outcomes and minimize the risk of complications or disease progression.

Practical Tip: When developing AI models for personalized treatment recommendations, it is important to ensure that the models are transparent, interpretable, and auditable. This means providing clear explanations of the data inputs, algorithms, and decision rules used by the models, as well as enabling clinicians and patients to review and challenge the recommendations if necessary. It is also important to validate the models on diverse patient populations and treatment scenarios, and to continuously monitor and update the models based on real-world clinical outcomes and feedback.

Here is an example of how to use a decision tree algorithm to generate personalized treatment recommendations based on patient features.

This example references a publicly available dataset (e.g., a synthetic dataset generated by Synthea) and includes a few best practices such as a train/test split, hyperparameter tuning, and basic interpretability. In reality, you would download and preprocess the Synthea CSV (or any other open clinical dataset) to reflect your actual use case.

1. Setup and data acquisition
   a. Generate or download a synthetic EHR dataset using Synthea.
   b. Export the generated data (e.g., patients.csv, encounters.csv, etc.) and merge the tables relevant to your task into a single CSV.
   c. Load the final CSV into a Pandas DataFrame.
2. Example code using a decision tree for personalized treatment

```
import pandas as pd
from sklearn.model_selection import train_test_split, GridSearchCV
from sklearn.tree import DecisionTreeClassifier, plot_tree
import matplotlib.pyplot as plt
```

```python
# 1. Load publicly available synthetic EHR data (e.g., Synthea CSV)
#    Assume columns might be: ['age', 'gender', 'biomarker1', 'biomarker2', 'treatment_response']
ehr_data = pd.read_csv('path/to/synthea_patient_data.csv')


# 2. Preprocess the data
def preprocess_data(df):
    # Example cleaning steps:
    #  - Handle missing values
    #  - Encode 'gender' as numeric
    #  - Convert biomarkers to numeric
    df = df.dropna(subset=['age', 'gender', 'biomarker1', 'biomarker2', 'treatment_response'])

    # Simple label encoding for gender
    df['gender'] = df['gender'].map({'Male': 0, 'Female': 1})

    # Ensure biomarkers are float
    df['biomarker1'] = df['biomarker1'].astype(float)
    df['biomarker2'] = df['biomarker2'].astype(float)

    return df

ehr_data = preprocess_data(ehr_data)
```

```python
# 3. Extract features and target variable
features = ['age', 'gender', 'biomarker1', 'biomarker2']
target = 'treatment_response'
X = ehr_data[features]
y = ehr_data[target]

# 4. Train/test split to evaluate performance
X_train, X_test, y_train, y_test = train_test_split(X, y,

test_size=0.2,

random_state=42,

stratify=y)

# 5. Hyperparameter tuning using grid search
param_grid = {
    'max_depth': [3, 5, 7, None],
    'min_samples_split': [2, 5, 10],
    'criterion': ['gini', 'entropy']
}
dt = DecisionTreeClassifier(random_state=42)
grid_search = GridSearchCV(dt, param_grid, cv=5, n_jobs=-1)
grid_search.fit(X_train, y_train)
```

```python
best_dt = grid_search.best_estimator_
print("Best hyperparameters:", grid_search.best_params_)
print("Training accuracy:", best_dt.score(X_train, y_train))
print("Test accuracy:", best_dt.score(X_test, y_test))


# 6. Generate personalized treatment recommendations for new patients
new_patient_data = pd.DataFrame({
    'age': [65],
    'gender': [0],   # 0=Male, 1=Female
    'biomarker1': [2.4],
    'biomarker2': [1.8]
})
recommendation = best_dt.predict(new_patient_data)[0]


# 7. Output personalized treatment recommendations
if recommendation == 'Responder':
    print('Recommended treatment: Drug A')
else:
    print('Recommended treatment: Drug B')


# (Optional) 8. Interpretability: Visualize the decision tree (requires matplotlib)
plt.figure(figsize=(12, 8))
plot_tree(best_dt, feature_names=features, class_names=best_dt.classes_, filled=True)
plt.show()
```

How This Approach Works

1. Publicly available or synthetic data
   - We reference a Synthea-generated CSV as an example. In practice, you would use the actual file path to your merged EHR or research dataset.
2. Preprocessing
   - This includes dropping rows with missing values for key fields, converting columns to numeric types, and encoding categorical variables (e.g., "Male" = 0 and "Female" = 1).
3. Train/test split and hyperparameter tuning
   - We split our data to evaluate the model's generalizability.
   - We use GridSearchCV to tune max_depth, min_samples_split, and criterion to find the best DecisionTreeClassifier configuration.
4. Personalized Treatment
   - We create a new patient DataFrame with example values (age, gender, biomarkers). The trained model predicts whether this patient is likely a "Responder" or not.
   - We then present treatment recommendations based on the predicted label.
5. Interpretability
   - We use plot_tree (from scikit-learn) to visualize the final decision tree. This can help clinicians and data scientists understand how the model is making decisions.

By coupling a decision tree model with proper data processing and hyperparameter tuning, you can generate more reliable, explainable recommendations for patient treatments, even when using publicly available or synthetic datasets as a starting point.

7.4 DRUG DISCOVERY AND REPURPOSING

AI is also being applied to the process of drug discovery and development, with the goal of identifying new therapeutic targets, optimizing drug design, and accelerating clinical trials. Traditional drug discovery is a lengthy, expensive, and often inefficient process, with high failure rates and long timelines from initial discovery to regulatory approval. AI has the potential to revolutionize this process by leveraging large-scale biological and chemical data to identify promising drug candidates and predict their safety and efficacy.

One application of AI in drug discovery is the use of machine learning algorithms to predict drug-target interactions and optimize drug candidate selection. By analyzing large datasets of chemical structures, protein sequences, and bioactivity assays, AI models can learn to identify patterns and relationships that are predictive of drug-target binding and therapeutic efficacy. This can help to narrow down the search space for potential drug candidates, and guide the design of new compounds with improved potency and selectivity.

AI is also utilized for drug discovery through the use of deep learning models to generate novel chemical structures with desired properties. By training on large datasets of known drugs and their chemical properties, deep learning models can learn to generate new molecules that are similar in structure and function to existing drugs, but with potentially improved pharmacological profiles. This can help to accelerate the discovery of new drug candidates, and reduce the need for expensive and time-consuming screening and optimization processes.

In addition to discovering new drugs, AI can also be used to identify new indications for existing drugs, a process known as drug repurposing. By analyzing large-scale clinical and genomic data from electronic health records and patient registries, AI models can identify patterns and relationships that suggest potential new uses for approved drugs. For example, by comparing the molecular signatures of diseases and drugs, AI models can identify drugs that may have therapeutic effects in diseases for which they were not originally developed. This can help to accelerate the development of new treatments for rare or neglected diseases, and reduce the costs and risks associated with de novo drug discovery.

Practical Tip: When developing AI models for drug discovery and repurposing, it is important to collaborate closely with domain experts in medicinal chemistry, pharmacology, and clinical medicine to ensure that the models are scientifically valid and clinically relevant. This may involve incorporating prior knowledge and constraints into the model design, such as known structure-activity relationships or pharmacokinetic properties, as well as validating the models on independent datasets and in experimental assays. It is also important to consider the potential risks and limitations of AI-generated drug candidates, such as the potential for off-target effects or adverse events, and to carefully evaluate their safety and efficacy in preclinical and clinical studies.

Load and Preprocess Clinical Data

This section is responsible for loading, preprocessing, and preparing clinical data for use in the model. Clinical data may include features such as patient demographics, lab results, and vital signs.

Installation and Data

Steps:

1. Load data

    - Load clinical data from a CSV file. Each row represents a patient, and columns include features (independent variables) and outcomes (dependent variables).

    ```
    data_df = pd.read_csv('path/to/clinical_data.csv')
    ```

2. Feature and target selection

    - feature_cols: columns representing independent variables (e.g., lab results, vitals, demographics)
    - target_cols: columns representing the dependent variable (e.g., clinical outcome like a risk score)

    ```
    feature_cols = [col for col in data_df.columns if 'feature_' in col]
    target_cols = ['outcome']
    X = data_df[feature_cols].values
    y = data_df[target_cols].values
    ```

3. Normalization

    - Normalize both features and outcomes for consistency, reducing the impact of variable scales. Standardization converts data to a mean of 0 and a standard deviation of 1.

    ```
    X = (X - np.mean(X, axis=0)) / np.std(X, axis=0)
    y = (y - np.mean(y)) / np.std(y)
    ```

Preprocessing ensures the data is clean and in a format that the model can efficiently learn from. Standardization is crucial for numerical stability in neural networks.

Variationl Autoencoder Architecture

The Variational Autoencoder (VAE) is a generative model used for dimensionality reduction. It learns a compressed representation of high-dimensional clinical data in a latent space while maintaining the ability to reconstruct the original input.

Encoder:

1. Inputs: Accepts patient data with input_dim features (e.g., 100 features like labs and vitals).

```
inputs = Input(shape=(input_dim,))
```

2. Hidden layers: Two dense layers (128 and 64 neurons) with ReLU activation, followed by Dropout layers to prevent overfitting

```
h = Dense(128, activation='relu')(inputs)
h = Dropout(0.2)(h)
h = Dense(64, activation='relu')(h)
h = Dropout(0.2)(h)
```

3. Latent variables: Mean (z_mean) and log-variance (z_log_var) define the latent space.

```
z_mean = Dense(latent_dim)(h)
z_log_var = Dense(latent_dim)(h)
```

4. Reparameterization trick: ensures differentiability during training by introducing a random noise component

```
def sampling(args):
    z_mean, z_log_var = args
    epsilon = tf.random.normal(shape=(tf.shape(z_mean)[0], latent_dim))
    return z_mean + tf.exp(0.5 * z_log_var) * epsilon
z = Lambda(sampling, output_shape=(latent_dim,))([z_mean, z_log_var])
```

Decoder:

1. Inputs: Accepts latent variables as input.

```
decoder_input = Input(shape=(latent_dim,))
```

2. Hidden layers: two dense layers with ReLU activation to reconstruct the original features

```
dh = Dense(64, activation='relu')(decoder_input)
dh = Dropout(0.2)(dh)
dh = Dense(128, activation='relu')(dh)
```

3. Output layer: outputs reconstructed patient features using sigmoid activation.

```
outputs = Dense(input_dim, activation='sigmoid')(dh)
```

Complete VAE Model

The VAE combines the encoder and decoder into a single model.

```
vae = Model(inputs, reconstructed, name='vae')
```

Train the VAE

This code trains the VAE to learn compressed representations of patient data in the latent space while ensuring the ability to reconstruct the original input.

1. Loss function:

    - Reconstruction Loss: Measures how well the decoder reconstructs the input.

```
reconstruction_loss = tf.reduce_mean(
    tf.keras.losses.mean_squared_error(inputs, reconstructed)
)
```

    - KL Divergence Loss: Regularizes the latent space to follow a Gaussian distribution.

```
kl_loss = -0.5 * tf.reduce_mean(
    1 + z_log_var - tf.square(z_mean) - tf.exp(z_log_var)
)
```

    - Total VAE Loss:

```
vae_loss = reconstruction_loss + kl_loss
```

2. Training process: The VAE learns to minimize the combined loss during training.

```
vae.compile(optimizer=Adam(learning_rate=0.001))
vae.fit(X, X, epochs=20, batch_size=32,
validation_split=0.1)
```

Why It Matters:

The Variational Autoencoder (VAE) approach is important because it compresses high-dimensional clinical data into a smaller, more tractable latent space, while still capturing essential features. This compression makes subsequent tasks—such as outcome prediction—both more efficient and potentially more accurate. In practice, clinicians and data scientists can extract latent representations from patient data via the VAE's encoder, then feed those lower-dimensional features into a separate model designed to predict outcomes like risk scores or treatment responses. By working from these learned embeddings, the prediction model can leverage the most clinically relevant information distilled by the VAE, reducing noise and computational overhead while preserving the critical signals that drive accurate forecasts.

1. Latent feature extraction: Use the encoder to extract the latent features from the original data.

```
encoder = Model(inputs, z_mean, name='encoder')
X_latent = encoder.predict(X)
```

2. Outcome prediction model: A feedforward neural network maps the latent features to clinical outcomes.

```
outcome_input = Input(shape=(latent_dim,))
h = Dense(64, activation='relu')(outcome_input)
h = Dropout(0.2)(h)
h = Dense(32, activation='relu')(h)
outcome_output = Dense(y.shape[1],
activation='linear')(h)
outcome_model = Model(outcome_input, outcome_output,
name='outcome_model')
```

3. Training: Train the outcome prediction model on the latent features.

```
outcome_model.compile(loss='mse', optimizer=Adam(learning_rate=0.001))
outcome_model.fit(X_latent, y, epochs=20, batch_size=32, validation_split=0.1)
```

Predicting clinical outcomes enables decision support by identifying at-risk patients or tailoring treatments based on patient features.

Predict and Interpret

Here, we apply the trained model to new patient data for clinical decision-making.

1. New patient data: Simulate or input real patient data, normalize it, and pass it through the encoder.

```
new_patient_data = np.random.normal(0, 1, (1, input_dim))
new_patient_data = (new_patient_data - np.mean(X, axis=0)) / np.std(X, axis=0)
latent_representation = encoder.predict(new_patient_data)
```

2. Outcome prediction: Use the trained outcome model to predict clinical outcomes for the patient.

```
predicted_outcome = outcome_model.predict(latent_representation)
print(f"Predicted clinical outcome: {predicted_outcome[0][0]}")
```

This step demonstrates how the trained model can be integrated into a clinical workflow to provide actionable insights.

Clinical Application Summary

- Dimensionality reduction: The VAE reduces complex clinical data into a compact representation.
- Outcome prediction: A downstream model uses the compact representation to predict clinical outcomes, enabling personalized medicine.
- Decision support: The framework can be integrated into Electronic Health Records (EHR) to assist clinicians in making evidence-based decisions.

Implementation Recommendations

1. Public dataset reference
   - Instead of a proprietary drug_data.npy, we reference an open dataset, such as ZINC or ChEMBL, which contain millions of SMILES strings and associated properties.
   - These datasets can be downloaded under permissive licenses for research.
2. Generative architecture (VAE)
   - We replace a feedforward network with a Variational Autoencoder, a common deep learning approach for molecule generation.
   - This allows us to sample from a latent distribution to generate novel chemical entities.
3. Property prediction
   - We optionally show how you might train a separate property prediction network on the latent vectors from the VAE. This helps steer the generation toward molecules with desired properties (sometimes done via multi-task learning or reinforcement learning approaches).
4. Use of RDKit
   - We include placeholders for RDKit usage—an industry-standard library for chemical informatics. In practice, you'll likely decode from a SMILES-based model or do a nearest-neighbor search in fingerprint space, then validate the resulting molecules for correctness.

5. Hyperparameter tuning and validation

    - We only show a skeleton for training. Real-world usage would include more thorough hyperparameter tuning (such as the number of layers, latent space dimension, and learning rate) and validation (e.g., checking generated structures are valid and meet property thresholds).

By using a VAE (or other generative model) with publicly available molecule data, you can learn a latent space that captures chemical structure and generate entirely new compounds that might have desired properties. This approach is the foundation of many de novo drug design pipelines in modern computational chemistry research.

This is just a simple example, but it illustrates the basic workflow of using deep learning models to generate novel chemical structures with desired properties. In real-world applications, there would be additional steps for data preprocessing, model architecture design, hyperparameter tuning, and validation on external datasets, as well as considerations for chemical feasibility, intellectual property, and regulatory compliance. Additionally, more advanced AI techniques, such as generative adversarial networks or reinforcement learning, could be used to generate more diverse and optimized chemical structures based on specific design criteria and objectives.

7.5 SURGICAL PLANNING AND ASSISTANCE

AI is also being integrated into surgical workflows to improve the precision, safety, and efficiency of surgical procedures. Surgical planning and execution often involve complex decision-making based on patient-specific anatomy, pathology, and risk factors, as well as real-time monitoring and adjustment based on intraoperative findings and events. AI can assist surgeons in these tasks by providing personalized guidance, real-time feedback, and intelligent automation.

One application of AI in surgical planning is the use of computer vision and machine learning algorithms to analyze preoperative images, such as CT or MRI scans, and create 3D models of patient-specific anatomy. These models can help surgeons to visualize and navigate complex anatomical structures, plan optimal surgical approaches and trajectories, and simulate different surgical scenarios. For example, AI algorithms can automatically segment and label different organs, vessels, and tumors from medical images, and provide quantitative measurements of their size, shape, and spatial relationships. This can help surgeons to identify critical structures to avoid, determine resection margins, and plan reconstructive procedures.

Another important application of AI in surgical assistance is the use of intelligent surgical robots and instruments that can augment the surgeon's skills and dexterity. These systems can use computer vision and machine learning algorithms to track the position and motion of surgical tools, provide real-time guidance and feedback, and automate repetitive or precision tasks. For example, AI algorithms can be used to control the movement of robotic arms and end-effectors, compensate for hand tremors or tissue deformation, and enforce safety constraints and virtual fixtures. This can help to reduce surgical errors, improve accuracy and consistency, and minimize tissue trauma and blood loss.

In addition to intraoperative assistance, AI can also be used to monitor and analyze surgical data in real-time, and provide decision support and early warning of potential complications. For example, AI algorithms can be used to analyze live video feeds from endoscopic cameras, and detect and highlight abnormal findings such as bleeding, inflammation, or perforation. AI can also be used to monitor patient vital signs, such as heart rate, blood pressure, and oxygenation, and alert the surgical team to any deviations from normal ranges. This can help to improve situational awareness, reduce cognitive load, and enable timely interventions and rescue procedures.

Practical Tip: When developing AI systems for surgical planning and assistance, it is important to involve surgeons and other clinical stakeholders throughout the design and validation process, to ensure that the systems are clinically relevant, user-friendly, and seamlessly integrated into existing workflows. This may involve conducting user studies and simulations to evaluate the usability, safety, and effectiveness of the AI systems, as well as providing adequate training and support for surgical teams to adopt and utilize the technology. It is also important to establish clear protocols and guidelines for the use of AI in surgery, including criteria for patient selection, informed consent, and post-operative follow-up, as well as mechanisms for monitoring and reporting adverse events and outcomes.

Below is an example that references a publicly available medical image dataset (the Medical Segmentation Decathlon) and demonstrates best practices such as reshaping, normalization, and applying the predicted segmentation mask. We assume you have a U-Net-like model trained on one of those public datasets (e.g., brain or liver segmentation). In reality, the exact input shape and data format will differ depending on your dataset (NIfTI, DICOM, and JPG). Here, we keep a simplified JPEG flow.

Setup and Data

1. Download a dataset from the Medical Segmentation Decathlon. These data are typically in NIfTI format; you can convert or extract slices to JPG/PNG for demonstration or use them in 3D directly with libraries such as NiBabel.
2. Train a segmentation model (e.g., a U-Net in Keras or PyTorch) on the labeled dataset.
3. Save your trained model (e.g., unet_segmentation_model.h5).

```
import numpy as np
import cv2
from keras.models import load_model
import tensorflow as tf


# ------------------------
# 1. Preprocessing Steps
# ------------------------

def preprocess_image(img, target_size=(256, 256)):
    """
    Example preprocessing function:
      - Convert to grayscale or keep 3 channels depending on your model.
      - Resize to match the model's expected input shape.
      - Normalize pixel values.
    """
    # If your model expects a single-channel image:
```

```python
    # img = cv2.cvtColor(img, cv2.COLOR_BGR2GRAY)  # convert to grayscale

    # Resize to model input size
    img = cv2.resize(img, target_size, interpolation=cv2.INTER_AREA)

    # Convert to float32 and normalize [0, 1]
    img = img.astype(np.float32) / 255.0

    # Expand dims to match (batch, height, width, channels) for Keras
    if len(img.shape) == 2:
        # single channel
        img = np.expand_dims(img, axis=-1)  # shape: (H, W, 1)
    return img

def apply_mask(original_img, mask):
    """
    Overlays the predicted mask onto the original image for visualization.
    Returns a color-coded overlay.
    """
    # Ensure original image is same size as mask
    h, w = mask.shape[:2]
    original_img = cv2.resize(original_img, (w, h))

    # Convert single-channel mask to color for overlay
    # Example: each label in mask gets a different
```

```python
    # For a binary mask (0 or 1), we'll just color label=1 in red:
    colored_mask = np.zeros_like(original_img)
    red_color = (0, 0, 255)  # BGR format
    colored_mask[mask == 1] = red_color

    # Blend images
    alpha = 0.5  # transparency factor
    overlay = cv2.addWeighted(original_img, 1 - alpha, colored_mask, alpha, 0)
    return overlay


def label_regions(segmented_img):
    """
    Example function to label connected components or regions in the segmentation.
    If you have multiple classes, you may do separate labeling or bounding boxes per class.
    Returns an image with text labels or bounding boxes.
    """
    # Convert to grayscale if needed
    gray = cv2.cvtColor(segmented_img, cv2.COLOR_BGR2GRAY)

    # Threshold to get binary mask (if not already)
    _, bin_mask = cv2.threshold(gray, 1, 255, cv2.THRESH_BINARY)
```

```python
    # Find connected components
    num_labels, labels_im = cv2.connectedComponents(bin_mask)

    # For each component, compute centroid and place label
    output = segmented_img.copy()
    for label_id in range(1, num_labels):  # skip label_id=0 (background)
        ys, xs = np.where(labels_im == label_id)
        cy, cx = int(ys.mean()), int(xs.mean())
        cv2.putText(output, f"Struct {label_id}", (cx, cy),
                    cv2.FONT_HERSHEY_SIMPLEX, 0.5, (255, 255, 0), 2)

    return output
# -----------------------
# 2. Inference with Trained Model
# -----------------------
# Load the original medical image (JPEG example)
original_image = cv2.imread('path/to/medical_image.jpg')
# Preprocess for model input
processed_image = preprocess_image(original_image)  # shape: (H, W, [C])
# Add batch dimension: (1, H, W, C)
input_for_model = np.expand_dims(processed_image, axis=0)
# Load the trained segmentation model (e.g., a U-Net)
```

```python
model = load_model('path/to/unet_segmentation_model.
h5', compile=False)


# Predict segmentation mask
# NOTE: shape could be (1, H, W, num_classes) if
multi-class
prediction = model.predict(input_for_model)


# If multi-class segmentation, take argmax across
channels:
mask = np.argmax(prediction, axis=-1)[0]  # shape: (H,
W)


# For binary segmentation (1 channel), do
thresholding:
# mask = (prediction[0, :, :, 0] > 0.5).astype(np.
uint8)


# ------------------------
# 3. Post-processing and Labeling
# ------------------------


# Overlay the segmentation mask
segmented_image = apply_mask(original_image, mask)


# Label the segmented regions
labeled_image = label_regions(segmented_image)
```

```
# Save the final labeled image
cv2.imwrite('path/to/labeled_image.jpg',
labeled_image)


print("Segmentation complete. Labeled image saved.")
```

How This Approach Works

1. Public dataset
   - We reference the Medical Segmentation Decathlon, which offers multiple organ/tissue segmentation tasks. You could also use other public datasets like BRATS (brain tumor), LiTS (liver tumor), or ACDC (cardiac MR segmentation).

2. Preprocessing
   - Medical images typically come in DICOM or NIfTI format, but in this snippet, we assume you have converted or extracted JPEG slices for demonstration.
   - We resize the image to match the trained model's input size, normalize pixel values, and expand dimensions for the batch.

3. Segmentation model
   - We assume you have a U-Net (or similar) trained on medical segmentation. The model returns a multi-channel output, where each channel corresponds to a specific class.
   - We apply argmax to get a label map for multiclass segmentation or threshold for binary segmentation.

4. Mask overlay and labeling
   - apply_mask converts the predicted mask into a color-coded overlay on the original image.
   - label_regions uses connected components to find distinct segmented structures, placing text labels at the centroid of each structure.

5. Saving output
   - We write the final labeled segmentation overlay to disk, which clinicians or researchers can review.

By utilizing a publicly available segmentation dataset (like those in the Medical Segmentation Decathlon) and a U-Net-style deep learning model, you can accurately segment anatomical structures in medical images, then label these regions for further analysis or clinical decision support.

This is just a simple example, but it illustrates the basic workflow of using computer vision techniques to segment and label anatomical structures from medical images. In real-world applications, there would be additional steps for data preprocessing, model architecture design, transfer learning, and post-processing, as well as considerations for image quality, anatomical variability, and clinical relevance. Additionally, more advanced AI techniques, such as 3D convolutional neural networks or adversarial learning, could be used to improve the accuracy and robustness of the segmentation and labeling process, and enable more sophisticated applications such as virtual reality simulation or augmented reality guidance.

7.5 READMISSION AND COMPLICATION PREDICTION

Hospital readmissions and complications are costly and often preventable events that can have a significant impact on patient outcomes and healthcare costs. AI can be used to predict the risk of these events and guide preventive interventions, such as targeted discharge planning, follow-up care, and patient education. By analyzing large-scale clinical and administrative data from electronic health records (EHRs) and other sources, AI models can identify patterns and risk factors that are predictive of readmissions and complications, and stratify patients based on their individual risk profiles.

One common approach to readmission and complication prediction is the use of machine learning algorithms, such as logistic regression, decision trees, or neural networks, to develop risk prediction models based on patient characteristics and clinical variables. These models can be trained on historical data from a specific hospital or health system, and validated on independent datasets to assess their accuracy and generalizability. For example, a readmission risk prediction model for heart failure patients might include variables such as age, gender, comorbidities, medication use, and previous hospitalizations, and generate a probability score for each patient's risk of readmission within 30 days of discharge.

Another important aspect of readmission and complication prediction is the incorporation of real-time data from wearable devices, remote monitoring systems, and patient-reported outcomes. By continuously monitoring patient vital signs, symptoms, and behaviors after discharge, AI models can detect early signs of deterioration or non-adherence, and trigger timely interventions by care teams. For example, an AI-powered remote monitoring system for post-surgical patients might use machine learning algorithms to analyze data from wearable sensors and patient-reported surveys, and alert clinicians to potential complications such as infection, bleeding, or pain.

In addition to predicting individual patient risk, AI can also be used to identify system-level factors and patterns that contribute to readmissions and complications, and guide quality improvement initiatives. For example, by analyzing EHR data across multiple hospitals or regions, AI models can identify variations in care processes, resource utilization, and outcomes, and benchmark performance against peer institutions. This can help to identify best practices and areas for improvement, and guide targeted interventions such as care standardization, process redesign, or staff training.

Practical Tip: When developing AI models for readmission and complication prediction, it is important to consider the potential biases and limitations of the data sources and algorithms used, and to validate the models on diverse patient populations and care settings. This may involve techniques such as data normalization, feature selection, and cross-validation, as well as engagement with clinical experts and patient stakeholders to ensure the face validity and actionability of the model outputs. It is also important to establish clear protocols and workflows for integrating the AI model outputs into clinical decision-making and care delivery, and to monitor and evaluate the impact of the models on patient outcomes and healthcare costs over time.

Below is an example showing how you might develop a logistic regression model for predicting hospital readmissions using publicly available EHR data (e.g., from MIMIC-IV on PhysioNet). This example includes basic preprocessing, cross-validation, and model evaluation steps to help ensure good model performance and generalizability. In practice, you would tailor the feature engineering and data cleaning processes to your specific dataset and clinical context.

Setup & Data Acquisition

1. Obtain a large EHR dataset such as MIMIC-IV (requires credentialing and acceptance of a data use agreement).

2. Extract or merge relevant tables (e.g., admissions, diagnoses, and procedures) to create a CSV for your predictive modeling task.
3. Load that CSV in Python (or read from a relational database) for subsequent data wrangling and model development.

Example Code for Logistic Regression

```
import pandas as pd
import numpy as np
from sklearn.linear_model import LogisticRegression
from sklearn.model_selection import train_test_split, cross_val_score
from sklearn.preprocessing import StandardScaler
from sklearn.metrics import accuracy_score, roc_auc_score, confusion_matrix


# -----------------------
# 1. Preprocessing Steps
# -----------------------


def preprocess_data(df):
    """
    Example preprocessing function for readmission modeling:
        - Drops rows with missing target variable
        - Encodes categorical variables (e.g., gender)
        - Scales numerical features
    """
```

```python
    # Drop rows where 'readmission' is NaN (if it exists in your dataset)
    df = df.dropna(subset=['readmission'])
    
    
    # Example: convert readmission to binary (0 or 1) if not already
    # e.g., readmission = 1 if readmitted within 30 days, else 0
    # If already binary, skip this step
    # df['readmission'] = (df['readmission'] < 30).astype(int)
    
    
    # Basic encoding for gender if it's categorical
    if 'gender' in df.columns:
        df['gender'] = df['gender'].map({'Male': 0, 'Female': 1})
    
    
    # Example numeric columns to scale (you'd customize this list)
    numeric_cols = ['age', 'num_comorbidities', 'length_of_stay', 'lab_value1', 'lab_value2']
    for col in numeric_cols:
        if col in df.columns:
            # fill missing numeric with median (if any)
            df[col] = df[col].fillna(df[col].median())
        else:
            # If not present in the dataset, skip
            numeric_cols.remove(col)
```

```python
    # Scale numeric columns
    scaler = StandardScaler()
    df[numeric_cols] = scaler.fit_transform(df[numeric_cols])

    # Drop rows with any remaining NaNs in relevant columns (optional)
    df = df.dropna(subset=numeric_cols)

    return df


# -----------------------
# 2. Load & Preprocess the EHR Data
# -----------------------
df = pd.read_csv('path/to/mimic_readmission_data.csv')
df = preprocess_data(df)

# Separate features (X) and target (y)
X = df.drop(['readmission'], axis=1)
y = df['readmission'].astype(int)   # ensure it's integer/binary


# -----------------------
# 3. Train/Test Split
# -----------------------
X_train, X_test, y_train, y_test = train_test_split(
    X, y, test_size=0.2, random_state=42, stratify=y
```

```python
# -----------------------
# 4. Train a Logistic Regression Model
# -----------------------
model = LogisticRegression(max_iter=1000)

# Optionally use cross-validation for a more robust performance estimate
cv_scores = cross_val_score(model, X_train, y_train, cv=5, scoring='roc_auc')
print(f"Mean CV AUC: {np.mean(cv_scores):.2f} +/- {np.std(cv_scores):.2f}")

# Fit the model on the entire training set
model.fit(X_train, y_train)


# -----------------------
# 5. Evaluate Model Performance
# -----------------------
y_pred = model.predict(X_test)
y_prob = model.predict_proba(X_test)[:, 1]  # Probability of class=1

accuracy = accuracy_score(y_test, y_pred)
auc = roc_auc_score(y_test, y_prob)
cm = confusion_matrix(y_test, y_pred)

print(f"Accuracy: {accuracy:.2f}")
```

```python
print(f"AUC: {auc:.2f}")
print("Confusion Matrix:")
print(cm)


# ------------------------
# 6. Predict Readmission Risk for New Patients
# ------------------------
# Example single-patient data (after applying same preprocessing steps)
new_patient_data = pd.DataFrame({
    'age': [65],
    'gender': [0],   # 0=Male, 1=Female
    'num_comorbidities': [3],
    'length_of_stay': [5],
    'lab_value1': [0.0],   # placeholder
    'lab_value2': [0.0],   # placeholder
})

# In practice, apply the same scaler/encoding used on training data.
# Here we assume the data is already preprocessed for demonstration.
readmission_risk = model.predict_proba(new_patient_data)[0][1]

print(f"Readmission risk: {readmission_risk:.2f}")
```

Important Refinements

1. Public data reference
   - We reference MIMIC-IV from PhysioNet as an example of a large, publicly available EHR dataset. The concepts apply equally to other datasets (e.g., from Kaggle or local hospital systems).

2. Preprocessing
   - We demonstrate a simplified approach: drop or impute missing values, encode categorical variables (gender), and scale numeric columns with StandardScaler.
   - Real-world pipelines often include more sophisticated cleaning, feature engineering (e.g., derived features from labs, vitals, or medication lists), and outlier detection.

3. Cross-validation
   - We use cross_val_score on the training set to get a more stable performance estimate (e.g., AUC). Then we retrain on the full training set for final deployment.

4. Confusion matrix and AUC
   - Accuracy alone can be misleading, especially if readmissions are relatively rare. We compute ROC AUC and a confusion matrix to better understand performance.

5. New patient predictions
   - We demonstrate how to predict readmission risk for a new patient, referencing the predict_proba method for probability estimates. Note that new data must follow the same preprocessing steps as the training data (e.g., scaling and encoding).

6. Practical considerations
   - Address potential biases (e.g., Does your dataset represent a diverse patient population?).
   - Update your model as practice patterns or patient demographics change over time (concept drift).
   - Collaborate with clinicians to define how predictions will be used and measured for real-world impact.

By using public EHR data (like MIMIC-IV), applying appropriate preprocessing steps, and performing cross-validation, you can develop a robust logistic regression model for hospital readmission prediction. Properly validated models can then support clinicians in identifying high-risk patients and implementing targeted interventions to reduce readmissions and improve outcomes.

This simple example illustrates the basic workflow of developing a machine learning model for predicting hospital readmissions based on EHR data. In real-world applications, there would be additional steps for data preprocessing, feature engineering, model selection, and hyperparameter tuning, as well as considerations for data privacy, model interpretability, and clinical integration. Additionally, more advanced AI techniques, such as ensemble learning, deep learning, or survival analysis, could be used to improve the accuracy and robustness of the readmission prediction models, and enable more personalized and timely interventions.

7.5.1 Chronic Disease Progression Modeling

Chronic diseases, such as diabetes, heart disease, and cancer, are the leading causes of death and disability worldwide. These conditions often involve complex and heterogeneous disease trajectories, with multiple risk factors, comorbidities, and treatment options that can vary over time and across individuals. AI can be used to model the progression of chronic diseases and identify opportunities for early intervention and personalized management, by leveraging large-scale clinical, behavioral, and molecular data from electronic health records, wearable devices, and other sources.

One application of AI in chronic disease progression modeling is the development of predictive models that can estimate the likelihood of disease onset, progression, and complications based on patient risk factors and biomarkers. These models can be trained on longitudinal data from patient cohorts or clinical trials, and validated on independent datasets to assess their accuracy and generalizability. For example, a diabetes progression model might include variables such as age, body mass index, blood glucose levels, and medication use, and predict the probability of developing complications such as retinopathy, neuropathy, or cardiovascular disease over a certain time horizon.

Another important aspect of chronic disease progression modeling is the identification of distinct patient subgroups or phenotypes that may have different disease trajectories, treatment responses, and outcomes. By using unsupervised learning techniques, such as clustering or latent variable modeling, AI can uncover hidden patterns and structures in patient data that may not be apparent from clinical observation alone. For example, a heart failure phenotyping model might identify subgroups of patients with different ejection fractions, biomarker profiles, and clinical presentations, and suggest tailored management strategies for each subgroup.

AI can also be used to simulate the impact of different interventions or treatment strategies on patient outcomes, using techniques such as Markov modeling, discrete event simulation, or reinforcement learning. These models can incorporate patient-specific data, as well as evidence from clinical guidelines and published literature, to generate personalized recommendations for lifestyle modifications, pharmacological therapies, or surgical procedures. For example, a reinforcement learning model for type 2 diabetes management might suggest optimal combinations and dosages of oral hypoglycemic agents and insulin based on a patient's glucose levels, side effect profiles, and treatment preferences.

Practical Tip: When developing AI models for chronic disease progression modeling, it is important to consider the potential limitations and biases of the data sources and algorithms used, and to validate the models on diverse patient populations and clinical settings. This may involve techniques such as data harmonization, missing data imputation, and transfer learning, as well as engagement with clinical experts and patient stakeholders to ensure the face validity and actionability of the model outputs. It is also important to establish clear protocols and workflows for integrating the AI model outputs into clinical decision-making and care delivery, and to monitor and evaluate the impact of the models on patient outcomes and healthcare costs over time.

Below is an example showing how you might construct a Markov model to simulate the progression of Chronic Kidney Disease (CKD). We reference publicly available datasets (e.g., from the United States Renal Data System (USRDS) or NHANES) as sources for transition probabilities and baseline characteristics. The snippet also demonstrates basic discounting for quality-adjusted life years (QALYs) and a time horizon over which the simulation runs. In practice, you would tailor the transition probabilities, states, costs, and QALY values (utilities) based on published literature or your specific data.

Setup and Data Sources

1. Public CKD data: Large-scale datasets like USRDS or NHANES can provide real-world estimates of CKD prevalence, progression rates (transitions between stages), and mortality.

2. Literature-based utilities: To compute QALYs, you can use health utility values for each CKD stage from peer-reviewed publications or health economic guidelines.

```python
import numpy as np
import pandas as pd

# 1. Define Markov states & transition probabilities
#    In practice, use real estimates from USRDS,
NHANES, or published studies
states = ['Stage 1', 'Stage 2', 'Stage 3', 'Stage 4',
'Stage 5', 'Death']
transition_probs = {
    'Stage 1': [0.7, 0.2, 0.1, 0.0, 0.0, 0.0],
    'Stage 2': [0.0, 0.6, 0.3, 0.1, 0.0, 0.0],
    'Stage 3': [0.0, 0.0, 0.5, 0.3, 0.1, 0.1],
    'Stage 4': [0.0, 0.0, 0.0, 0.4, 0.4, 0.2],
    'Stage 5': [0.0, 0.0, 0.0, 0.0, 0.5, 0.5],
    'Death':   [0.0, 0.0, 0.0, 0.0, 0.0, 1.0]
}

# 2. Patient characteristics & treatment (example data)
patient_data = pd.DataFrame({
    'age': [65],
    'diabetes': [True],
    'hypertension': [True],
    'egfr': [45],
    'acr': [200],
```

```python
        'treatment': ['ACE inhibitor']
})


# 3. Time horizon & discount rate
time_horizon = 10
discount_rate = 0.03


# 4. Initialize the Markov model
#    Start in 'Stage 1' with probability=1 (illustrative).
#    For a real scenario, start in an appropriate stage based on egfr, etc.
state_vector = [1, 0, 0, 0, 0, 0]
n_states = len(states)
state_matrix = np.zeros((n_states, time_horizon))
state_matrix[:, 0] = state_vector


# 5. Simulate Markov transitions
for t in range(1, time_horizon):
    current_state_index = np.argmax(state_vector)
    current_state_name = states[current_state_index]
    # Next state probabilities come from the row for current_state_name
    transition_vector = transition_probs[current_state_name]
    # Compute new state distribution
    state_vector = np.dot(state_vector, transition_vector)
    state_matrix[:, t] = state_vector
```

```python
# 6. Calculate discounted QALYs
#    Example utility weights by CKD stage (placeholder):
utilities = np.array([0.9, 0.8, 0.7, 0.6, 0.5])  # for stages 1-5
# Exclude the 'Death' state from QALY calculation
qaly_matrix = state_matrix[:-1, :] * utilities[:, np.newaxis]


# Discount factor for each year
discount_factors = 1 / ((1 + discount_rate) ** np.arange(time_horizon))


# Multiply by discount factors and sum
qalys = np.sum(qaly_matrix * discount_factors)


print(f"Discounted QALYs over {time_horizon} years: {qalys:.2f}")
```

How This Example Works

1. States and transitions
   - We define six states: Stage 1—5 for CKD progression, plus Death.
   - The transition_probs dictionary provides row vectors for each state, indicating the probabilities of moving to every possible next state in one cycle (e.g., 1 year).

2. Patient characteristics
   - The simple patient_data DataFrame includes key CKD risk factors (e.g., diabetes, hypertension, eGFR, and ACR). In a more advanced model, you could adjust the transition probabilities dynamically based on a patient's profile or interventions.

3. Markov simulation
    - We maintain a state_vector representing the patient's probability distribution across states (e.g., a 1 for "Stage 1" at t=0).
    - Each cycle, we dot this vector with the transition probabilities to compute the next year's probabilities.
    - We store these results in state_matrix for future analysis.
4. Utility weights and discounting
    - We assign a utility for each CKD stage (e.g., 0.9 for Stage 1, 0.8 for Stage 2, and 0.5 for Stage 5). These are placeholders; real values come from published quality-of-life studies or validated instruments.
    - We exclude Death from QALY calculation.
    - Each year's utility contribution is discounted using a discount_rate (3% in this example). The sum of discounted utilities gives the total QALYs.
5. Expansion and customization
    - Calibrate or validate the transition probabilities using real-world CKD data (e.g., USRDS or NHANES).
    - Incorporate treatment effects (e.g., ACE inhibitors, dialysis, and transplant) by modifying transitions dynamically or splitting states further (e.g., "Stage 5 with dialysis" and "Stage 5 with transplant").
    - Add costs (medical, treatment, and hospitalization) for a cost-effectiveness analysis, discounting them similarly to QALYs.

Key Takeaways

- Public data: By utilizing USRDS or NHANES, you can ground your transition probabilities and utility scores in real-world evidence.
- Model validation: Validate or calibrate the Markov model by comparing projected CKD progression rates with observed outcomes in your dataset.
- Clinical use: This type of model helps estimate long-term outcomes (QALYs, costs) under various treatment strategies (e.g., ACE inhibitors vs. ARBs vs. new therapies).
- Customization: Expand the model to incorporate patient heterogeneity, comorbidities, and dynamic transitions as risk factors change over time.

By following these steps and referencing publicly available data sources, you can build a more robust, evidence-based Markov model for CKD progression and analyze the impact of different interventions on patient outcomes and healthcare costs.

7.5.2 Social Determinants of Health Analysis

Social determinants of health (SDOH) are the economic and social conditions that influence individual and group differences in health status. These include factors such as income, education, employment, housing, transportation, social support, and access to healthcare, which can have a profound impact on health outcomes and disparities. AI can be used to analyze SDOH data from various sources, such as electronic health records, social services databases, and public health surveillance systems, to identify patterns and risk factors associated with adverse health outcomes, and guide interventions to address them.

One application of AI in SDOH analysis is the use of natural language processing (NLP) and text mining techniques to extract SDOH information from unstructured data sources, such as clinical notes, social work assessments, and patient-reported outcomes. By using NLP algorithms to identify and classify SDOH-related concepts and sentiments, such as mentions of financial hardship, food insecurity, or transportation barriers, AI can help to capture a more comprehensive and nuanced picture of patients' social and environmental contexts, and inform targeted interventions and referrals.

Another important application of AI in SDOH analysis is the development of predictive models that can estimate the impact of SDOH factors on health outcomes and healthcare utilization. By using machine learning algorithms to analyze large datasets of SDOH and health data, AI can identify complex interactions and nonlinear relationships between social and clinical variables, and generate risk scores or stratification models that can guide population health management and resource allocation. For example, a predictive model for hospital readmissions might incorporate SDOH variables such as housing instability, social isolation, and medication affordability, in addition to clinical factors such as comorbidities and previous hospitalizations.

In addition to predicting individual-level outcomes, AI can also be used to analyze SDOH data at a population or community level, to identify geographic and demographic disparities in health and social determinants, and guide public health interventions and policy decisions. By using geospatial analysis and data visualization techniques, AI can map the distribution of SDOH factors across different neighborhoods or regions, and identify "hotspots" or "coldspots" of health inequities that may require targeted investments or community partnerships. For example, an AI-powered SDOH dashboard might display indicators such as poverty rates, educational attainment, and access to green space for different census tracts or zip codes, and allow users to explore the relationships between these indicators and health outcomes such as life expectancy or chronic disease prevalence.

PRACTICAL TIP *When developing AI models for SDOH analysis, it is important to engage diverse stakeholders, including patients, community members, social service providers, and public health experts, to ensure that the models are culturally sensitive, ethically sound, and actionable. This may involve using participatory design methods, such as focus groups or community advisory boards, to gather input on the relevant SDOH factors and desired outcomes, and to co-create solutions that are tailored to local needs and assets. It is also important to consider the potential biases and limitations of the data sources and algorithms used, and to validate the models on diverse populations and settings, to ensure their generalizability and equity.*

Below is a refined code example demonstrating how you might use NLP techniques (via spaCy) to extract Social Determinants of Health (SDOH) information (e.g., housing, transportation, substance use, food insecurity) from clinical notes. This example uses rule-based entity matching (via EntityRuler), but in a production environment, you may consider training a specialized model or using existing pretrained scispaCy or clinical NLP models.

NOTE *This snippet is illustrative. In a real-world scenario, you would gather a labeled dataset (e.g., from MIMIC-IV or similar) where portions of text corresponding to SDOH have been annotated, and then either train or fine-tune an NLP model to recognize these entities.*

Setup

```
Install spaCy and scispaCy (optional)
pip install spacy scispacy
pip install https://s3-us-west-2.amazonaws.com/ai2-s2-scispacy/releases/en_core_sci_md-0.5.1.tar.gz
```

TIP *You can use en_core_sci_md (scispaCy) for better performance on biomedical text, or simply en_core_web_sm if you want to keep it lightweight.*

Code

```
import spacy
from spacy.pipeline import EntityRuler


# -----------------------
# 1. Load a Base NLP Model
# -----------------------
# For general English text, you might use 'en_core_web_sm'.
# For biomedical/clinical text, consider 'en_core_sci_md' (scispaCy).
nlp = spacy.load("en_core_web_sm")


# -----------------------
# 2. Create an EntityRuler for SDOH Terms
# -----------------------
ruler = EntityRuler(nlp, overwrite_ents=True)

# Define some example patterns for SDOH extraction
# In practice, you'd expand these or make them more sophisticated (regex, etc.)
patterns = [
    {
        "label": "SDOH_HOUSING",
        "pattern": [{"LOWER": {"IN": ["homeless", "housing", "eviction"]}}],
    },
```

```python
    {
        "label": "SDOH_FOOD_INSECURITY",
        "pattern": [{"LOWER": {"IN": ["food", "hunger", "insecurity"]}}],
    },
    {
        "label": "SDOH_TRANSPORTATION",
        "pattern": [{"LOWER": {"IN": ["transportation", "bus", "ride", "car", "vehicle"]}}],
    },
    {
        "label": "SDOH_SUBSTANCE_USE",
        "pattern": [{"LOWER": {"IN": ["alcohol", "tobacco", "smoking", "drugs", "substance"]}}],
    },
    {
        "label": "SDOH_SOCIAL_SUPPORT",
        "pattern": [{"LOWER": {"IN": ["support", "family", "caregiver", "spouse"]}}],
    },
]

ruler.add_patterns(patterns)
nlp.add_pipe(ruler, before="ner")   # Insert rules before the built-in NER component


# -----------------------
# 3. Example Clinical Note with SDOH Mentions
```

```python
# ------------------------
clinical_note = """
Patient is a 52-year-old male with history of hypertension and chronic back pain.
He is currently living in a shelter and reports difficulty obtaining regular meals.
He also reports smoking tobacco 1 pack/day.
Transportation is a challenge because he doesn't have a car.
However, he has strong family support.
"""

# ------------------------
# 4. NLP Processing & Extraction
# ------------------------
doc = nlp(clinical_note)

sdoh_entities = []
for ent in doc.ents:
    if ent.label_.startswith("SDOH"):
        sdoh_entities.append((ent.text, ent.label_))

# ------------------------
# 5. Output the Extracted SDOH Information
# ------------------------
print("Extracted SDOH Entities:")
for text, label in sdoh_entities:
    print(f"  - {label}: {text}")
```

How This Works

1. EntityRuler setup
   - We create an EntityRuler object that matches SDOH keywords in lowercase. Each pattern is assigned a label (e.g., SDOH_HOUSING, SDOH_FOOD_INSECURITY).

2. Example clinical note
   - We include references to housing, food, smoking, transportation, and social support to simulate typical SDOH mentions.

3. Rule-Based Extraction
   - When doc = nlp(clinical_note) runs, the text is tokenized, and spaCy's pipeline processes it. Our EntityRuler flags matches to the patterns we defined, producing labeled entities.

4. Output
   - We print a list of recognized SDOH phrases and their labels (e.g., ("shelter", "SDOH_HOUSING")).

Enhancements and Next Steps

1. Regex and advanced patterns
   - Instead of simple keyword matching, you can define regex or more nuanced patterns to handle phrases like "lack of stable housing" or "missed meals."

2. Machine learning approach
   - A robust method is to train a NER model on annotated clinical notes for SDOH-specific labels. Tools like scispaCy or BioClinicalBERT can be fine-tuned for better domain-specific performance.

3. Context and negation
   - You may need to handle negations or context (e.g., "denies homelessness") using libraries like negspaCy or additional spaCy components.

4. Structured output
   - Convert extracted SDOH data into structured formats (e.g., FHIR resources) for downstream analytics or population health dashboards.

5. Ethical and privacy considerations
   - SDOH data is often sensitive. Ensure your data governance and HIPAA compliance processes are in place if working with real patient data.

By defining custom rule-based patterns or training a specialized model, you can use spaCy (or scispaCy) to automatically extract key SDOH factors from clinical notes, thereby helping healthcare teams identify social needs that may impact patient outcomes.

This is a rather simple example, but it illustrates the basic workflow of using NLP techniques to extract SDOH information from unstructured clinical text. In real-world applications, there would be additional steps for data cleaning, preprocessing, and post-processing, as well as considerations for data privacy, model performance, and clinical relevance. Additionally, more advanced NLP techniques, such as sentiment analysis, topic modeling, or deep learning, could be used to improve the accuracy and granularity of the SDOH extraction, and enable more sophisticated applications such as automatic risk stratification or social prescribing.

7.6 TECHNICAL BARRIERS AND INTEROPERABILITY ISSUES

Implementing AI in clinical workflows requires overcoming a number of technical barriers and challenges, particularly related to data quality, standardization, and interoperability. Healthcare data is often fragmented, heterogeneous, and inconsistent across different systems and settings, which can limit the ability to develop and deploy robust and generalizable AI models. Some key technical barriers and interoperability issues include the following:

1. Data quality and completeness: Clinical data is often incomplete, inaccurate, or inconsistent, due to factors such as human error, data entry variations, and system limitations. Missing or erroneous data can bias AI models and lead to incorrect or suboptimal predictions and recommendations. Ensuring data quality and completeness requires rigorous data validation, cleaning, and imputation techniques, as well as standardized data capture and documentation practices.

2. Data standardization and normalization: Clinical data is often represented in different formats, terminologies, and coding systems across different healthcare organizations and settings. This lack of standardization

can hinder data integration, analysis, and model development, as well as limit the portability and scalability of AI solutions. Standardizing and normalizing data requires the use of common data models, ontologies, and vocabularies, such as FHIR, SNOMED-CT, and LOINC, as well as data mapping and transformation tools.

3. Data privacy and security: Healthcare data contains sensitive and protected health information (PHI), which is subject to strict privacy and security regulations, such as HIPAA and GDPR. Developing and deploying AI models requires careful consideration of data access, use, and disclosure policies, as well as appropriate technical safeguards, such as data encryption, anonymization, and access controls. Balancing data privacy and utility can be challenging, and may require novel techniques such as federated learning or differential privacy.

4. Infrastructure and computational resources: Developing and deploying AI models requires significant computational resources, such as high-performance computing clusters, storage systems, and network bandwidth. Many healthcare organizations may lack the necessary infrastructure and expertise to support large-scale AI initiatives, particularly for resource-intensive applications such as medical imaging or natural language processing. Cloud computing and software-as-a-service (SaaS) platforms can help to alleviate some of these challenges, but may also introduce additional concerns related to data security, vendor lock-in, and cost.

5. Model interpretability and explainability: Many AI models, particularly deep learning models, are often considered "black boxes," meaning that their internal logic and decision-making processes are not easily understandable or explainable to end-users. This lack of interpretability can limit the trust and adoption of AI solutions by healthcare providers and patients, as well as raise ethical and legal concerns related to accountability and liability. Developing interpretable and explainable AI models requires the use of techniques such as feature importance analysis, rule extraction, and counterfactual reasoning, as well as clear communication and visualization of model outputs.

Practical Tip: To address technical barriers and interoperability issues, it is important to adopt a collaborative and iterative approach to AI development and deployment, involving close coordination between data scientists, IT staff, and clinical stakeholders. This may involve establishing data governance and stewardship processes, such as data quality assessment and monitoring, data standards and harmonization, and data access and use policies. It may also involve using agile and human-centered design methodologies, such as rapid prototyping, user testing, and continuous improvement, to ensure that AI solutions are technically feasible, clinically relevant, and user-friendly.

Here is an example of how to use the FHIR (Fast Healthcare Interoperability Resources) standard to normalize and exchange clinical data for AI model development:

We reference a publicly available dataset (again using MIMIC-IV) as a conceptual source for patient, encounter, and observation data. In reality, you would need to map your specific table schemas or CSV columns to the relevant FHIR resources (e.g., Patient, Encounter, and Observation) using the python-fhirclient library or similar tooling.

The snippet below is illustrative. In production, you would also handle authorization, resource validation, error-checking, and batch uploads to a FHIR server (such as HAPI FHIR or Google Cloud Healthcare FHIR API).

Install the FHIR client library in Python:

```
pip install fhirclient
```

Obtain a publicly available EHR dataset (e.g., MIMIC-IV from PhysioNet). Then create or identify a FHIR server (local or cloud-hosted) to store and retrieve resources, e.g., HAPI FHIR installed locally or a hosted service.

```
import pandas as pd
from fhirclient import client
from fhirclient.models.patient import Patient
from fhirclient.models.humanname import HumanName
from fhirclient.models.fhirdate import FHIRDate
from fhirclient.models.observation import Observation
from fhirclient.models.quantity import Quantity
from fhirclient.models.encounter import Encounter
from fhirclient.models.fhirreference import FHIRReference
```

```python
# ------------------------
# 1. FHIR Client Settings
# ------------------------
# Replace 'http://localhost:8080/fhir' with the base URL of your FHIR server.
settings = {
    'app_id': 'my_fhir_app',
    'api_base': 'http://localhost:8080/fhir'  # e.g., local HAPI FHIR server
}
fhir_client = client.FHIRClient(settings=settings)


# ------------------------
# 2. Load & Process Example EHR Data
# ------------------------
# Hypothetical CSV from MIMIC-like data with columns: patient_id, name, dob, lab_code, lab_value, encounter_id
df = pd.read_csv('path/to/mimic_ehr_subset.csv')


# For demonstration, assume each row is a lab observation for a single patient encounter.
# We will convert selected columns to FHIR resources:
#    - Patient: basic demographics (Name, DOB)
#    - Encounter: ID or reference
#    - Observation: lab results
# In reality, you might also create Condition, MedicationRequest, etc.
```

```python
# ----------------------
# 3. Create FHIR Resources
# ----------------------
patients_created = {}
encounters_created = {}

for idx, row in df.iterrows():
    patient_id = str(row['patient_id'])
    encounter_id = str(row['encounter_id'])
    lab_code = row['lab_code']
    lab_value = row['lab_value']
    name_str = row['name']
    dob_str = row['dob']

    # -- (A) Construct/Update a Patient resource if not already created
    if patient_id not in patients_created:
        patient = Patient()
        # Name
        hn = HumanName()
        hn.text = name_str  # e.g., "John Doe"
        patient.name = [hn]
        # BirthDate
        patient.birthDate = FHIRDate(dob_str)  # e.g., "1965-07-01"
        # Assign an ID for local tracking (could be ephemeral)
```

```python
        patient.id = f"patient-{patient_id}"

        # Optionally POST to FHIR server to create Patient
        # saved_patient = patient.create(fhir_client.server)

        patients_created[patient_id] = patient

    # -- (B) Construct/Update Encounter resource if not already created
    if encounter_id not in encounters_created:
        encounter = Encounter()
        encounter.id = f"encounter-{encounter_id}"
        # Reference to Patient
        encounter.subject = FHIRReference()
        encounter.subject.reference = f"Patient/patient-{patient_id}"

        # Optionally POST to create Encounter
        # saved_encounter = encounter.create(fhir_client.server)

        encounters_created[encounter_id] = encounter

    # -- (C) Construct an Observation resource for the lab
    observation = Observation()
    observation.id = f"obs-{idx}"  # or a unique ID
    observation.status = "final"
```

```python
    # CodeableConcept representation for the lab code
    observation.code = {
        "coding": [{
            "system": "http://loinc.org",
            "code": lab_code,
            "display": "Example Lab Test"
        }]
    }

    # Reference to Patient
    observation.subject = FHIRReference()
    observation.subject.reference = f"Patient/patient-{patient_id}"

    # Reference to Encounter
    observation.encounter = FHIRReference()
    observation.encounter.reference = f"Encounter/encounter-{encounter_id}"

    # Value as a Quantity
    observation.valueQuantity = Quantity({
        "value": float(lab_value),
        "unit": "mg/dL",   # Example unit
        "system": "http://unitsofmeasure.org",
        "code": "mg/dL"
    })
```

```
        # Optionally POST Observation to the server
        # saved_obs = observation.create(fhir_client.
server)


    # ----------------------
    # 4. Summarize or Upload Resources
    # ----------------------
    print(f"Created {len(patients_created)} Patient
resources.")
    print(f"Created {len(encounters_created)} Encounter
resources.")
    print(f"Total Observations processed: {len(df)}")


    # (Optional) If you haven't posted them in the loop,
you could do it here:
    # for patient in patients_created.values():
    #     patient.create(fhir_client.server)
    # for encounter in encounters_created.values():
    #     encounter.create(fhir_client.server)
    #     ...
```

Explanation and Important Steps

1. FHIR client configuration
   - We use python-fhirclient to interact with a FHIR server. The settings dictionary includes the base URL of your FHIR endpoint (e.g., a local or cloud-hosted HAPI FHIR server).
2. Data acquisition
   - We conceptually load a CSV from MIMIC-IV (or any EHR data source). In reality, you would do a more detailed mapping between your dataset's columns and the appropriate FHIR elements.

3. Creating FHIR resources
   - Patient: basic demographics (e.g., name, and date of birth)
   - Encounter: establishes the clinical context of the patient's visit, including timing, location, and reason for the healthcare interaction
   - Observation: represents lab tests, vitals, or other measured values
4. Referencing resources
   - Each Encounter references a Patient (encounter.subject.reference = "Patient/...").
   - Observation references both the Patient and the Encounter. This linkage is crucial to maintain correct relationships in the FHIR ecosystem.
5. Storing data in the FHIR server
   - You can call resource.create(fhir_client.server) to POST the resource to the FHIR server, or resource.update(...) to update an existing resource. The snippet above shows placeholders where you could do so.
6. Why use FHIR for AI?
   - By standardizing your clinical data in FHIR, you can more easily share data across systems, ensure semantic consistency, and facilitate broader interoperability.
   - For AI model development, you can pull relevant resources (like Observations for labs, Conditions for diagnoses, and MedicationRequests for meds) into a consistent format from different healthcare systems that expose FHIR APIs.

Additional Considerations

- Authentication and security: Production FHIR servers typically require OAuth2 or other mechanisms. You would configure these in settings or handle them externally.

- Batch upserts: For large datasets, consider using batch/transaction endpoints in FHIR to efficiently upload or update multiple resources in one request.

- Validation and profiles: FHIR has resource profiles to ensure standard compliance. You can run the created resources against a FHIR validator to confirm correctness.

- Resource extensions: If your data has fields not covered by base FHIR resource definitions, you can use extensions or define custom StructureDefinitions.

- Analytics: Once data are in FHIR format, they can be retrieved for analytics or AI using RESTful queries (e.g., GET /Observation?code=loinc|1234-5&patient=...). This allows seamless integration of the data into your machine learning pipelines.

By converting your EHR data into FHIR resources, you promote interoperability and ensure consistent data definitions across healthcare systems. This normalized format allows AI developers to more easily share, merge, and analyze clinical datasets, ultimately improving model development and deployment. More advanced FHIR features, such as resource versioning, terminology binding, and clinical reasoning, could be used to enable more sophisticated and interoperable AI solutions.

7.7 WORKFLOW DISRUPTION AND USER ACCEPTANCE

Integrating AI into clinical workflows also requires careful consideration of the potential impacts on existing processes, roles, and relationships. AI tools and systems can disrupt established workflows and practices, and may be met with resistance or skepticism by healthcare providers and patients. Some challenges related to workflow disruption and user acceptance include the following:

1. Workflow compatibility and integration: AI tools and systems must be designed and implemented in a way that is compatible with existing clinical workflows and processes. This requires a deep understanding of the clinical context, user needs, and data flows, as well as the ability to seamlessly integrate AI outputs and recommendations into electronic health records (EHRs), clinical decision support systems (CDSS), and other health IT platforms. Poorly designed or integrated AI tools can create additional workload, confusion, or errors for healthcare providers, and may be abandoned or require a work-around.

2. User trust and adoption: Healthcare providers and patients must trust and accept AI tools and recommendations in order for them to be effectively used and acted upon. This requires clear communication and transparency about the AI model's purpose, performance, and limitations, as well as the ability to explain and justify its outputs in a way that is understandable and meaningful to users. AI tools that are perceived as opaque, biased,

or unreliable may be met with skepticism or resistance, and may undermine user confidence and engagement.

3. Liability and accountability: The use of AI in healthcare raises complex questions related to liability and accountability for AI-based decisions and outcomes. Healthcare providers may be hesitant to rely on AI recommendations if they are uncertain about their legal and ethical responsibilities, or if they fear being held liable for AI errors or adverse events. Clear policies and guidelines are needed to define the roles and responsibilities of AI developers, users, and other stakeholders, as well as to establish mechanisms for monitoring, auditing, and redressing AI-related harms.

4. Skill and knowledge gaps: The effective use of AI in healthcare requires new skills and knowledge on the part of healthcare providers and other users. This includes the ability to interpret and apply AI outputs, to understand the strengths and limitations of different AI approaches, and to communicate and collaborate with AI developers and data scientists. Addressing these skill and knowledge gaps requires targeted education and training programs, as well as ongoing support and resources for users to stay up-to-date with the latest AI developments and best practices.

5. Unintended consequences and ethical concerns: The use of AI in healthcare can also have unintended consequences and raise ethical concerns related to issues such as privacy, fairness, and autonomy. For example, AI tools that rely on sensitive health data may pose risks to patient privacy and confidentiality, while AI algorithms that are biased or discriminatory may exacerbate health disparities and undermine trust in the healthcare system.

Proactively identifying and mitigating these risks requires ongoing monitoring and evaluation of AI tools and their impacts, as well as engagement with diverse stakeholders to ensure that AI is developed and used in an ethical and socially responsible manner.

Practical Tip: To address workflow disruption and user acceptance challenges, it is important to adopt a human-centered and participatory approach to AI development and implementation. This may involve using user-centered design methods, such as workflow analysis, usability testing, and co-design workshops, to ensure that AI tools are tailored to user needs and preferences. It may also involve establishing clear communication and feedback channels between AI developers, users, and other stakeholders, as well as providing ongoing training, support, and resources to help users effectively integrate AI into their workflows and decision-making processes.

7.7.1 Cost and Resource Constraints

Implementing AI in healthcare also requires significant financial and organizational resources, which can be a barrier for many healthcare systems, particularly those serving underserved or resource-limited populations. Developing, deploying, and maintaining AI tools and systems can be costly and time-consuming, and may require specialized expertise and infrastructure that is not readily available or affordable. Some challenges related to cost and resource constraints include the following:

1. Upfront development and infrastructure costs: Developing AI models and systems requires significant upfront investments in data collection, processing, and annotation, as well as in computing hardware, software, and storage. These costs can be particularly high for data-intensive applications such as medical imaging or natural language processing, which may require large datasets, complex algorithms, and specialized equipment. Additionally, many healthcare organizations may lack the necessary IT infrastructure and expertise to support AI development and deployment, and may need to invest in new systems, personnel, and training.

2. Ongoing maintenance and update costs: AI models and systems also require ongoing maintenance, monitoring, and updating to ensure their performance, security, and relevance. This includes tasks such as data refreshing, model retraining, software patching, and hardware upgrades, which can be costly and time-consuming. Additionally, as clinical knowledge and practices evolve, AI models may need to be adapted or replaced to reflect the latest evidence and guidelines, which can further increase maintenance costs and complexity.

3. Reimbursement and payment models: Another challenge is the lack of clear and consistent reimbursement and payment models for AI-based healthcare services. Many current payment systems are based on fee-for-service or volume-based models, which may not adequately incentivize or reward the use of AI for quality improvement, cost reduction, or population health management. Additionally, there may be uncertainty or variability in how payers and regulators evaluate and reimburse AI-based services, which can create financial risks and disincentives for healthcare organizations to adopt AI.

4. Equitable access and adoption: Cost and resource constraints can also exacerbate existing disparities in healthcare access and outcomes,

particularly for underserved or disadvantaged populations. Healthcare organizations serving these populations may have limited resources and capacity to invest in AI development and adoption, and may be less likely to benefit from AI-based innovations and improvements. Additionally, AI models and systems that are developed and trained on data from privileged or majority populations may not generalize well to diverse or marginalized populations, and may even perpetuate or amplify biases and inequities.

5. Opportunity costs and trade-offs: Finally, investing in AI development and adoption may come at the expense of other healthcare priorities and needs, such as staffing, infrastructure, or patient care services. Healthcare organizations may need to make difficult trade-offs and decisions about how to allocate limited resources and budgets, and may face competing demands and pressures from different stakeholders and constituencies. Balancing these trade-offs and ensuring that AI investments align with organizational values and goals can be challenging, and may require careful planning, communication, and stakeholder engagement.

Practical Tip: To address cost and resource constraints, healthcare organizations can explore various strategies and approaches, such as the following:

- partnering with academic institutions, technology vendors, or other healthcare organizations to share data, expertise, and resources for AI development and deployment
- utilizing open-source software, cloud computing, and other cost-effective technologies to reduce upfront and ongoing costs of AI infrastructure and maintenance
- developing and testing AI models and systems incrementally and iteratively, starting with small-scale pilots and proofs-of-concept, and scaling up based on demonstrated value and impact
- engaging with payers, policymakers, and other stakeholders to advocate for more flexible and value-based reimbursement and payment models that incentivize and support AI adoption and innovation
- prioritizing AI investments and initiatives based on strategic goals, patient needs, and health equity considerations, and involving diverse stakeholders in the planning and decision-making process

7.7.2 Bias and Fairness in AI Models

One of the ethical concerns around the use of AI in healthcare is the potential for bias and unfairness in AI models and recommendations. AI models are only as good as the data they are trained on, and if that data is biased or unrepresentative, the resulting models may perpetuate or even amplify those biases. This can lead to AI systems that discriminate against certain groups of patients based on factors such as race, ethnicity, gender, age, or socioeconomic status, and that exacerbate existing health disparities and inequities.

There are several potential sources of bias in AI models for healthcare, including the following:

1. Biased data: AI models are often trained on historical data from electronic health records, claims databases, or clinical trials, which may contain biases and disparities that reflect past and current inequities in healthcare access, quality, and outcomes. For example, if certain populations are underrepresented or misrepresented in the training data, the resulting models may not perform well or may make incorrect or harmful predictions for those populations.

2. Biased algorithms: Even if the training data is unbiased, the algorithms and models used to analyze that data may introduce biases through their design, architecture, or optimization process. For example, some algorithms may be more sensitive to certain features or patterns in the data, or may optimize for certain outcomes or metrics that are not aligned with fairness or equity goals.

3. Biased implementation: Even if the data and algorithms are unbiased, the way in which AI models are implemented and used in clinical practice may introduce biases through human decision-making, communication, or interpretation. For example, if clinicians over-rely on AI recommendations without considering other relevant factors, or if they apply AI outputs differently to different patient populations, they may perpetuate or exacerbate biases and disparities.

To address these concerns, it is important to develop and implement rigorous methods for detecting and mitigating bias in AI models for healthcare, such as the following:

1. Data bias audits: regularly assessing and monitoring the quality, representativeness, and fairness of the data used to train and validate AI models, and taking steps to correct or compensate for any identified biases or gaps

2. Algorithm bias testing: systematically evaluating the performance and fairness of AI algorithms and models across different subgroups and populations, using metrics such as demographic parity, equalized odds, or individual fairness, and adjusting the models as needed to ensure equitable outcomes

3. Inclusive and participatory design: involving diverse stakeholders, including patients, community members, and domain experts, in the design, development, and evaluation of AI models, to ensure that they are aligned with the needs, values, and experiences of the populations they aim to serve

4. Transparency and accountability: providing clear and accessible information about the data, algorithms, and assumptions underlying AI models, and establishing mechanisms for monitoring, auditing, and redressing any biases or harms that may arise from their use

Practical Tip: When developing and implementing AI models for healthcare, it is important to proactively assess and address potential biases and fairness issues at every stage of the process, from data collection and preprocessing to model training and evaluation to deployment and monitoring. This requires close collaboration between AI developers, healthcare providers, patients, and other stakeholders, as well as ongoing vigilance and responsiveness to any identified biases or disparities. Some specific strategies and best practices for mitigating bias in healthcare AI include the following:

- using diverse and representative datasets for model training and validation, and applying techniques such as data augmentation, reweighting, or stratification to correct for any imbalances or gaps
- selecting and optimizing algorithms and models based on fairness and equity metrics, such as equal opportunity, demographic parity, or individual fairness, in addition to traditional performance metrics such as accuracy or AUC
- conducting regular bias audits and fairness assessments of AI models and their outputs, using both quantitative and qualitative methods, and involving diverse stakeholders in the evaluation and interpretation of the results
- providing clear and accessible explanations of AI model decisions and recommendations, using techniques such as feature importance, counterfactual analysis, or rule extraction, and enabling users to challenge or override the model outputs when appropriate

- establishing clear policies and procedures for monitoring, reporting, and mitigating any biases or harms that may arise from the use of AI in healthcare, and ensuring that there are adequate resources and incentives for ongoing quality improvement and equity promotion

Aequitas is an open-source library that provides a set of tools for auditing and mitigating bias in machine learning models, based on various fairness metrics and visualizations. This toolkit is particularly valuable in healthcare applications, where biased algorithms could contribute to healthcare disparities and inequitable treatment.

```
import pandas as pd
from aequitas.group import Group
from aequitas.bias import Bias
from aequitas.fairness import Fairness

# Load the data and define the relevant columns
data = pd.read_csv('path/to/data.csv')
label_col = 'readmission'
protected_attr_cols = ['race', 'gender', 'age']
score_col = 'risk_score'

# Create a Group object and calculate the bias metrics
group = Group()
group.fit(data, label_col, protected_attr_cols, score_col)

# Create a Bias object and calculate the fairness metrics
bias = Bias()
bias.fit(group)
```

```python
# Create a Fairness object and calculate the fairness metrics
fairness = Fairness()
fairness.fit(group)


# Plot the bias and fairness metrics
bias.plot()
fairness.plot()


# Mitigate the bias and unfairness using different techniques
from aequitas.preprocessing import preprocess_input_df
from sklearn.linear_model import LogisticRegression
from sklearn.metrics import roc_auc_score


# Reweight the data to balance the protected attribute groups
df_reweighted = preprocess_input_df(data, protected_attr_cols)


# Train a new model on the reweighted data
model = LogisticRegression()
model.fit(df_reweighted[['risk_factor_1', 'risk_factor_2']], df_reweighted[label_col])


# Evaluate the new model's performance and fairness
df_reweighted['risk_score'] = model.predict_proba(df_reweighted[['risk_factor_1', 'risk_factor_2']])[:, 1]
```

```
group_reweighted = Group()
group_reweighted.fit(df_reweighted, label_col,
protected_attr_cols, score_col)
bias_reweighted = Bias()
bias_reweighted.fit(group_reweighted)
fairness_reweighted = Fairness()
fairness_reweighted.fit(group_reweighted)

print(f"Original AUC: {roc_auc_score(data[label_col],
data[score_col]):.3f}")
print(f"Reweighted AUC: {roc_auc_
score(df_reweighted[label_col],
df_reweighted[score_col]):.3f}")
bias_reweighted.plot()
fairness_reweighted.plot()
```

In this example, we first load a healthcare dataset and define the relevant columns, including the label (readmission), protected attributes (race, gender, and age), and risk score. We then use the Aequitas toolkit to assess the bias and fairness of the original risk score model.

We create a Group object to calculate the bias metrics across the protected attribute groups, a Bias object to calculate the fairness metrics based on the bias metrics, and a Fairness object to calculate additional fairness metrics such as demographic parity and equalized odds.

We plot the bias and fairness metrics using the built-in visualization functions, which show the disparities and inequities in the model's performance and outcomes across the protected attribute groups.

To mitigate the identified biases and unfairness, we apply a preprocessing technique called reweighting, which adjusts the sample weights of the data points to balance the protected attribute groups. We then train a new logistic regression model on the reweighted data, and evaluate its performance and fairness using the same Aequitas metrics and visualizations. Then compare the AUC scores of the original and reweighted models, and plot the bias and fairness metrics of the reweighted model to assess the effectiveness of the mitigation technique.

This example demonstrates how Aequitas can be used to systematically assess and mitigate bias and unfairness in healthcare AI models, using a combination of quantitative metrics, visualizations, and preprocessing techniques. However, it is important to note that bias mitigation is an ongoing process that requires careful monitoring, iteration, and stakeholder engagement, and that no single technique or tool can completely eliminate bias or ensure perfect fairness. It is also important to consider the potential trade-offs and unintended consequences of different mitigation strategies, and to involve diverse perspectives and expertise in the evaluation and decision-making process.

7.8 EXPLAINABILITY AND TRANSPARENCY

Another ethical concern around AI in healthcare is the need for explainability and transparency in AI decision-making. Many AI models, particularly those based on deep learning or complex algorithms, are often considered *black boxes*, meaning that their internal logic and reasoning are not easily understandable or interpretable by human users. This lack of explainability can make it difficult for clinicians and patients to trust and rely on AI recommendations, and can raise concerns about accountability, liability, and informed consent. Let's examine why explainability and transparency are important for AI in healthcare and clinical decision support.

Clinical decision support. In healthcare, clinicians must understand and justify AI-driven recommendations in order to effectively integrate them with their own expertise and to communicate them clearly to patients and colleagues. If the reasoning behind an AI model's output is unclear, clinicians may hesitate to rely on it, or they may misinterpret or misuse the guidance. Explainable AI helps build trust by offering transparent insights into how the model arrived at its recommendations, making it more likely that clinicians will adopt AI suggestions in a responsible and appropriate manner.

Patient autonomy and informed consent. Patients have the right to understand the factors that influence AI-based healthcare decisions and to make well-informed choices about their care. By revealing how AI models generate recommendations—such as the variables they rely on or the rationale behind specific outputs—explainable AI ensures that patients can better evaluate their options according to personal values, preferences, and circumstances.

This openness supports a more collaborative decision-making process and respects each patient's autonomy.

Accountability and liability. When AI models guide high-stakes decisions in healthcare—for instance, diagnosing conditions, selecting treatments, or allocating resources—it is essential to establish a clear chain of responsibility for any negative outcomes. By shedding light on the internal workings of AI systems and pinpointing the causes of errors or miscalculations, explainable AI helps assign appropriate accountability to developers, providers, or administrators. This transparency not only facilitates redress for patients but also promotes safer and more reliable AI deployments in clinical settings.

Regulatory and legal compliance. Healthcare AI is regulated by entities such as the FDA in the United States and must also comply with broader data protection laws like the EU's General Data Protection Regulation (GDPR). Explainable AI assists in meeting these requirements by clearly documenting the data, algorithms, and processes used within the model, as well as providing avenues for meaningful human oversight and control. Through this clarity, stakeholders can more readily demonstrate compliance, fostering confidence among regulators, clinicians, and patients alike.

To address these concerns, there are several approaches and techniques for developing and implementing explainable AI in healthcare.

Feature importance and attribution. Identifying and visualizing the features or variables that contribute most to an AI model's outputs is crucial for understanding why a model behaves the way it does. Techniques such as permutation importance, SHAP values, and saliency maps highlight these influential inputs, enabling clinicians and other stakeholders to see which factors are driving AI recommendations. By relating these factors to known domain knowledge or clinical guidelines, users can validate or question the AI's reasoning, ultimately promoting trust and more informed decision-making.

Rule extraction and decision trees. Even when using complex AI models like deep neural networks, it is often helpful to have more transparent, human-interpretable representations. Rule extraction or decision trees derived through techniques like decision tree induction, rule-based learning, or knowledge distillation can yield simpler models that clearly outline the logic behind predictions. Although there may be a trade-off in terms of reduced accuracy or granularity, these interpretable models can be especially useful in high-stakes settings, such as healthcare, where transparent decision-making is paramount.

Counterfactual explanations and case-based reasoning. Another powerful way to understand AI model outputs is to explore how they might change if certain inputs were modified or if different conditions applied. Methods like counterfactual generation, adversarial perturbation, or case-based reasoning simulate hypothetical scenarios, revealing the boundaries and limitations of a model's understanding. By examining these "what-if" situations, users can gain insight into potential edge cases, failure modes, and the robustness of AI models, which is critical for safe deployment in real-world applications.

Natural language explanations and dialogue. Finally, presenting AI explanations in plain language—such as through chatbots, language models, or interactive visualizations—can greatly enhance the accessibility and usability of AI systems. When users can engage in a dialogue, asking questions or seeking clarifications, they are more likely to trust and adopt AI-based recommendations. A conversational approach fosters a continuous feedback loop, allowing users to refine model outputs or flag inaccuracies in real time.

Practical Tip: When developing explainable AI for healthcare, it is important to consider the diverse needs, preferences, and abilities of different user groups, such as clinicians, patients, regulators, or researchers. This may require using multiple explanation techniques or modalities, and tailoring them to the specific context and purpose of the AI application. It is also important to involve users in the design, evaluation, and refinement of AI explanations, using participatory and human-centered methods such as user testing, co-design, or stakeholder workshops.

Some specific strategies and best practices for implementing explainable AI in healthcare include the following:

- selecting and optimizing AI models and algorithms that are inherently more interpretable or explainable, such as decision trees, rule-based systems, or generalized additive models, in addition to more complex or opaque models such as deep neural networks
- providing multiple levels or types of explanations, from high-level summaries to detailed technical information, and enabling users to drill down or navigate between them based on their needs and interests
- using domain-specific terminology, visualizations, and examples to make AI explanations more relevant and meaningful to healthcare users, and aligning them with clinical guidelines, protocols, and best practices

- conducting rigorous evaluations and validations of AI explanations, using both quantitative metrics (e.g., faithfulness, stability, sensitivity) and qualitative feedback from users and stakeholders, and iteratively improving them based on the results
- documenting and communicating the methods, assumptions, and limitations of AI explanations, using standardized frameworks such as model cards, datasheets, or ethical principles, and making them publicly available for scrutiny and feedback

Here is an example of how to use the SHAP (SHapley Additive exPlanations) library to generate feature importance explanations for a healthcare AI model.

```
import pandas as pd
import shap
from sklearn.ensemble import RandomForestClassifier
from sklearn.model_selection import train_test_split
from sklearn.metrics import classification_report, roc_auc_score
import matplotlib.pyplot as plt


# Load the data and ensure no missing values
data = pd.read_csv('path/to/data.csv')
if data.isnull().sum().any():
    print("Data contains missing values. Handle them before proceeding.")
    data = data.fillna(data.median())  # Replace missing values with the median


# Convert categorical variables to numeric (if necessary)
if data['gender'].dtype == 'object':
```

```python
    data['gender'] = data['gender'].map({'Male': 0, 'Female': 1})

# Split into features and target
X = data[['age', 'gender', 'bmi', 'bp', 'cholesterol']]
y = data['diabetes']

# Train-test split
X_train, X_test, y_train, y_test = train_test_split(X, y, test_size=0.2, random_state=42)

# Train a Random Forest classifier
model = RandomForestClassifier(n_estimators=100, random_state=42)
model.fit(X_train, y_train)

# Evaluate model performance
y_pred = model.predict(X_test)
y_proba = model.predict_proba(X_test)[:, 1]
print(classification_report(y_test, y_pred))
print("ROC-AUC:", roc_auc_score(y_test, y_proba))

# Generate SHAP explanations for the test set
explainer = shap.TreeExplainer(model)

# Subsample test data for faster SHAP computation (if test set is large)
```

```python
subsample = X_test.sample(100, random_state=42) if
len(X_test) > 100 else X_test
shap_values = explainer.shap_values(subsample)


# Visualize the SHAP explanations for a single
instance
shap.initjs()
instance_index = 0  # Modify to view other instances
shap.force_plot(
    explainer.expected_value[1],
    shap_values[1][instance_index, :],
    subsample.iloc[instance_index, :]
)

# Save the force plot for the instance
force_plot_html = shap.force_plot(
    explainer.expected_value[1],
    shap_values[1][instance_index, :],
    subsample.iloc[instance_index, :]
)
shap.save_html("force_plot.html", force_plot_html)

# Visualize the SHAP summary plot for all instances
shap.summary_plot(shap_values[1], subsample)
plt.savefig('shap_summary_plot.png')
```

```python
# Visualize SHAP dependence plots for each feature
for feature in X.columns:
    shap.dependence_plot(feature, shap_values[1], subsample)
    plt.savefig(f'shap_dependence_plot_{feature}.png')

# Additional feature importance visualization
feature_importances = pd.DataFrame({
    'Feature': X.columns,
    'Importance': model.feature_importances_
}).sort_values(by='Importance', ascending=False)
print(feature_importances)

# Plot feature importance for visual understanding
plt.figure(figsize=(10, 6))
plt.barh(feature_importances['Feature'], feature_importances['Importance'])
plt.title('Feature Importance')
plt.xlabel('Importance')
plt.ylabel('Feature')
plt.gca().invert_yaxis()
plt.savefig('feature_importance.png')
plt.show()
```

In this example, we first loaded a healthcare dataset and split it into features (age, gender, BMI, blood pressure, and cholesterol) and target (diabetes) variables.

We then train a random forest classifier on the data, using an 80/20 train/test split and default hyperparameters.

To generate SHAP explanations for the trained model, we create a TreeExplainer object and pass it the model and the test set features. The TreeExplainer is a specific implementation of SHAP for tree-based models such as random forests or gradient boosting machines.

We can then visualize the SHAP explanations for individual instances, using the force_plot function. This shows the positive and negative contributions of each feature to the model's output for a given instance, as well as the expected value (i.e., the average output of the model across all instances).

We can also visualize the SHAP explanations for all instances, using the summary_plot function. This shows the overall importance and direction of each feature across the entire dataset, as well as the distribution of SHAP values for each feature.

Finally, we can visualize the SHAP explanations for each feature individually, using the dependence_plot function. This shows how the model's output changes as a function of each feature, while accounting for the interactions and correlations with other features.

These visualizations provide a powerful and intuitive way to explain the behavior and reasoning of complex AI models, and to help users understand and trust their outputs. However, it is important to note that SHAP explanations are not perfect or complete, and may have limitations or biases depending on the specific model, data, and problem domain. It is also important to use SHAP in combination with other explanation techniques and user feedback, and to critically evaluate and validate the explanations before using them in practice.

7.8.1 Patient Privacy and Data Ownership

The use of AI in healthcare also raises important questions around patient privacy and data ownership. AI models often require large amounts of patient data for training and validation, including sensitive information such as medical records, genetic profiles, and biometric data. This data is subject to strict privacy and security regulations, such as the Health Insurance Portability and Accountability Act (HIPAA) in the US, or the General Data Protection Regulation (GDPR) in the EU, which govern how it can be collected, used, and shared by healthcare providers, researchers, and technology companies.

Challenges and Risks for Patient Privacy and Data Ownership in AI

The rapid growth and increasing complexity of AI in healthcare have introduced serious concerns about safeguarding patient privacy and clarifying data ownership. As patient information becomes a more valuable resource for clinical decision-making and research, it also becomes more susceptible to various forms of misuse. From potential data breaches to questions about who truly controls health information, understanding these challenges is important to building trustworthy and ethical AI solutions.

Data Breaches and Cyberattacks

In the digital age, patient data is both a critical asset and a prime target for malicious actors. With large volumes of personal health information now stored and transmitted electronically, cybercriminals can gain unauthorized access, steal sensitive data, or even manipulate AI systems used in clinical settings. Because AI models often rely on extensive and intricate datasets, they are especially attractive targets for cyberattacks. A breach of this scale can have severe implications for patient privacy and healthcare outcomes, emphasizing the urgent need for robust security measures.

Secondary Use and Commercialization

Patient data initially gathered for legitimate medical purposes may find itself repurposed in ways patients never intended or consented to. For instance, organizations might share or sell data to third parties for marketing, advertising, or profiling. Such secondary usage without explicit patient permission erodes trust between patients and healthcare providers. It also undermines the expectation of confidentiality, leading to increased skepticism about how health data is handled in research and commercial ventures.

Algorithmic Bias and Discrimination

The quality and representativeness of the data used to train AI models directly affect their reliability and fairness. When those datasets contain imbalances or biases related to demographics, such as race, gender, or socioeconomic status, AI systems risk perpetuating existing healthcare disparities. This can manifest in biased treatment recommendations or diagnostic inaccuracies for certain groups, further widening health inequities. Vigilantly assessing and mitigating bias in AI models is therefore paramount to ensuring equitable care.

Lack of Transparency and Control

Despite being the principal stakeholders in their own care, patients often have minimal insight into how their data is utilized or how AI influences their medical decisions. They may be unaware that their information has been shared for secondary purposes or remains part of ongoing AI development. Without clear mechanisms for patients to review, opt out, or influence how their data is processed, questions of autonomy and informed consent inevitably arise. Building robust frameworks for data governance and transparency can help reinforce patient trust and preserve the integrity of AI-driven healthcare.

Principles and Best Practices for Responsible AI Data Use in Healthcare

To address the challenges associated with patient privacy, data ownership, and ethical considerations, healthcare organizations and researchers can adopt a set of guiding principles. By ensuring that these principles are woven into policy, technology, and clinical workflows, stakeholders can uphold patient trust, mitigate risk, and promote more equitable and beneficial uses of AI. Below are several best practices that have been widely proposed and implemented.

Data Minimization and Purpose Limitation

One of the core strategies for reducing privacy risks is to collect and use only the minimum amount of patient data required for a given AI application or research question. By confining data usage to its original purpose and limiting long-term retention, organizations can shrink the window of opportunity for data breaches or misuse. This approach not only honors patient privacy but also aligns with emerging data protection laws that emphasize proportionality and necessity.

Data Security and Access Controls

Robust technical and organizational measures are vital for safeguarding patient information. Techniques such as data encryption (both in transit and at rest), role-based access controls, comprehensive audit logs, and data loss prevention solutions can significantly reduce the likelihood of unauthorized access or tampering. Coupled with regular security assessments and staff training, these measures help maintain the confidentiality, integrity, and availability of patient data throughout its life cycle.

Transparency and Consent

Clearly communicating how and why patient data is collected, processed, and shared is essential for building trust. Organizations must ensure that patients receive accessible, plain-language explanations of the benefits and risks of AI-driven healthcare, as well as their rights regarding data usage. Seeking explicit, informed consent empowers patients to make choices aligned with their personal preferences, thereby reinforcing accountability and fostering a collaborative relationship between patients and providers.

Data Governance and Stewardship

A structured approach to data management ensures that patient information is employed responsibly and ethically. Effective governance frameworks should detail policies on data quality checks, bias and fairness assessments, and ongoing monitoring of AI systems. They should also provide transparent channels for patient feedback, complaint resolution, and redress. By establishing strong oversight mechanisms, healthcare entities can maintain public confidence in how patient data is harnessed for AI innovations.

Data Ownership and Control

Finally, many experts advocate for models that extend greater rights and benefits to patients regarding the use of their health information. Approaches like personal health records, data cooperatives, or data trusts allow patients to actively manage the flow of their data and potentially share in the gains produced by AI research or commercialization. Granting individuals more meaningful control can help shift the balance of power in healthcare, spurring patient-centered innovation and promoting ethical data ecosystems.

Practical Tip: When developing or deploying AI models that use patient data, it is important to engage early and often with patients, caregivers, and patient advocacy groups, to understand their perspectives, concerns, and preferences regarding data privacy and ownership. This can be done through various methods, such as surveys, interviews, focus groups, or deliberative forums, which can provide valuable insights and feedback on issues such as data sharing, consent, and benefits.

It is also important to work closely with legal, ethical, and regulatory experts, to ensure that the collection, use, and sharing of patient data for AI complies with applicable laws, standards, and best practices. This may require conducting privacy impact assessments, data protection impact assessments, or ethical reviews, and implementing appropriate safeguards and oversight mechanisms.

Finally, it is important to be transparent and accountable about the use of patient data for AI, and to provide clear and accessible information to patients about their rights and options regarding their data. This can include providing patients with access to their own data, the ability to correct or delete their data, or the ability to opt-out of certain data uses or disclosures.

Here is an example of how to use the PyHealth library to implement privacy-preserving machine learning on healthcare data.

```
import pandas as pd
from pyhealth.models import PrivacyPreservingLogisticRegression
from pyhealth.data import HealthDataLoader
from pyhealth.evaluation import auc, f1_score
from pyhealth.privacy import DifferentialPrivacy
# Load and preprocess the healthcare data
data_loader = HealthDataLoader(
    data_path='path/to/data.csv',
    label_col='diagnosis',
    feature_cols=['age', 'gender', 'bmi', 'bp', 'cholesterol'],
    categorical_cols=['gender'],
    normalize=True,  # Ensures features are normalized for better model performance
    test_size=0.2,  # 80-20 train-test split
    random_state=42  # Ensures reproducibility
)

# Load the train and test datasets
```

```python
X_train, X_test, y_train, y_test = data_loader.load_data()
# Initialize the privacy-preserving logistic regression model
model = PrivacyPreservingLogisticRegression(
    epsilon=1.0,  # Privacy budget parameter (smaller values mean stronger privacy)
    l2_norm_clip=1.0,  # Clip gradients to this norm to enforce differential privacy
    noise_multiplier=1.0,  # Noise added for differential privacy
    num_microbatches=10  # Number of microbatches for gradient computation
)
# Train the model with differential privacy
print("Training the privacy-preserving logistic regression model...")
model.fit(X_train, y_train)
# Evaluate the model performance on the test set
y_pred = model.predict(X_test)
print("\nModel Evaluation Metrics:")
print(f"AUC: {auc(y_test, y_pred):.3f}")
print(f"F1 Score: {f1_score(y_test, y_pred):.3f}")

# Evaluate the model's privacy budget
privacy_engine = DifferentialPrivacy()
epsilon, delta = privacy_engine.get_privacy_spent(model)
print("\nPrivacy Metrics:")
```

```python
print(f"Epsilon (Privacy Budget): {epsilon:.2f}")
print(f"Delta: {delta:.6f}")


# Additional Feature Importance (Optional)
if hasattr(model, "feature_importances_"):
    feature_importances = pd.DataFrame({
        'Feature': data_loader.feature_cols,
        'Importance': model.feature_importances_
    }).sort_values(by='Importance', ascending=False)
    print("\nFeature Importances:")
    print(feature_importances)


# Save the model (Optional)
model.save('privacy_preserving_logistic_regression_model.pkl')
```

In this example, we first load and preprocess a healthcare dataset using the HealthDataLoader class from PyHealth. This class provides a convenient way to load and split the data into training and testing sets, as well as to normalize and encode the features.

We then initialize a PrivacyPreservingLogisticRegression model, which is a variant of logistic regression that incorporates differential privacy techniques to protect the privacy of the training data. Differential privacy is a mathematical framework that allows for the release of aggregate statistics or machine learning models, while limiting the risk of individual data points being identified or reconstructed from the outputs.

Parameters in Privacy-Preserving Logistic Regression

A PrivacyPreservingLogisticRegression model balances the need for accurate predictions against the imperative to protect individual data points. One important parameter is *epsilon*, also known as the *privacy budget*, which defines the acceptable amount of privacy loss. Smaller epsilon values confer stronger privacy guarantees but can reduce model accuracy because they constrain the model's capacity to learn fine-grained details from the training data. Another important parameter, l2_norm_clip, places an upper bound on the L2 norm of the model's weights, thus limiting how sensitively the model can respond to outliers or unique data entries. Although smaller values improve privacy by preventing large weight updates tied to individual records, they can also narrow the model's representational power.

Additionally, the noise_multiplier determines how much Gaussian noise is added to the gradients during training, directly influencing the level of differential privacy. A larger noise multiplier bolsters privacy protection but may degrade performance by masking some of the useful signal in the data. Finally, num_microbatches represents the number of smaller batches into which the training dataset is split. Processing data in these microbatches allows for a tighter privacy analysis and can make more efficient use of the privacy budget. However, increasing the number of microbatches tends to prolong training time, illustrating the constant trade-off between stronger privacy, computational cost, and predictive accuracy.

We train the model on the training set using the fit method, which applies the differential privacy techniques to the model weights and outputs. We evaluate the model performance on the test set using the predict method, and calculate the AUC and F1 score metrics using the corresponding functions from PyHealth.

You can also evaluate the model privacy on the training set using the DifferentialPrivacy class from PyHealth, which calculates the actual privacy loss (epsilon) and failure probability (delta) of the model, based on the composition of the differential privacy mechanisms used during training.

This example demonstrates how PyHealth can be used to implement privacy-preserving machine learning on healthcare data, using differential privacy techniques. However, it is important to note that differential privacy is not a panacea for data privacy, and may have limitations or trade-offs depending on the specific data, model, and application. It is also important to consider other privacy-enhancing techniques, such as secure multi-party computation, homomorphic encryption, or federated learning, and to engage with patients, regulators, and other stakeholders to ensure that the privacy and ethical implications of the AI model are properly addressed.

CHAPTER 8

CASE STUDIES IN CLINICAL DECISION SUPPORT SYSTEMS

8.1 LEARNING FROM THE PIONEERS: REAL-WORLD EXAMPLES

Clinical decision support systems (CDSSs) have emerged as a critical tool in modern healthcare, aiding clinicians in making informed, evidence-based decisions at the point of care. By leveraging vast amounts of clinical data, medical knowledge, and patient-specific information, CDSSs can provide timely and accurate recommendations, alerts, and reminders to support various aspects of patient care, from diagnosis and treatment to monitoring and follow-up (Sutton et al., 2020).

The integration of AI techniques, such as machine learning, natural language processing, and knowledge representation, has significantly enhanced the capabilities of CDSS. AI-powered CDSSs can process and analyze complex, heterogeneous data from multiple sources, including electronic health records, medical imaging, genomic data, and patient-generated health data, to uncover hidden patterns, predict outcomes, and personalize recommendations (Jiang et al., 2017). By continuously learning from new data and feedback, these intelligent systems can adapt to changing clinical contexts and individual patient needs, improving their accuracy and relevance over time.

However, the development and implementation of AI-powered CDSSs in real-world clinical settings pose significant challenges, ranging from data quality and interoperability issues to concerns about transparency, bias, and user acceptance. To realize the full potential of these systems, it is crucial to learn from the experiences of early adopters and innovators who have successfully navigated these challenges and demonstrated tangible impact on patient care and outcomes.

In this chapter, we consider four case studies showcasing the diverse applications and benefits of an AI-powered CDSS across different clinical domains and settings. These real-world examples illustrate the important technical, organizational, and human factors that influence the success of CDSS implementations, as well as the lessons learned and future directions for advancing the field. By providing a practical perspective on the current state and potential of AI in clinical decision support, we aim to inspire and guide healthcare organizations, researchers, and practitioners in harnessing the power of AI to improve the quality, efficiency, and personalization of patient care.

8.2 CASE STUDY 1: AI-POWERED DIAGNOSTIC DECISION SUPPORT FOR RARE DISEASES

Rare diseases present a unique set of challenges for healthcare professionals, as their low prevalence, diverse symptomatology, and often limited available data make timely and accurate diagnosis difficult. Patients with rare diseases frequently experience prolonged diagnostic interactions, going from one specialist to another, undergoing countless tests, and often receiving misdiagnoses before arriving at a correct diagnosis. This delay in diagnosis can lead to inappropriate or delayed treatments, poorer health outcomes, reduced quality of life, and increased healthcare costs (Shen et al., 2019).

In response to these challenges, a pioneering healthcare technology company has developed an innovative AI-powered CDSS specifically designed to assist clinicians in diagnosing rare diseases. This groundbreaking system leverages state-of-the-art machine learning algorithms and comprehensive knowledge graphs to integrate and analyze data from a wide array of sources, providing clinicians with evidence-based, patient-specific diagnostic recommendations.

At the heart of this CDSS is a sophisticated data integration and processing pipeline that collects and harmonizes information from multiple disparate sources. Using advanced natural language processing (NLP) techniques, the system extracts pertinent data points from unstructured clinical notes, such as patient-reported symptoms, physical examination findings, and relevant family history. This unstructured data is then combined with structured information from electronic health records (EHRs), including patient demographics, laboratory test results, and imaging findings, creating a holistic and comprehensive understanding of each patient (Shen et al., 2019).

One of the innovations of this CDSS is its use of a dynamically updated rare disease knowledge graph. This knowledge graph is a comprehensive, machine-readable representation of the complex interrelationships between rare diseases, their signs and symptoms, risk factors, and associated genes, curated from the latest scientific literature, expert clinical knowledge, and patient registries (Shen et al., 2019). By continuously incorporating new research findings and real-world evidence, the knowledge graph ensures that the CDSS always provides the most current and accurate information to support diagnostic decision-making.

When a patient with a suspected rare disease is entered into the system, the CDSS employs a number of cutting-edge machine learning algorithms, including random forests, gradient boosting machines, and deep learning models, to analyze the patient's integrated clinical profile against the rare disease knowledge graph. This process generates a ranked list of the most likely differential diagnoses, each accompanied by a confidence score, detailed explanations of the supporting evidence, and links to relevant literature and clinical guidelines (Shen et al., 2019). The system also provides recommendations for additional diagnostic tests or specialist referrals to help confirm or rule out the suggested diagnoses.

The development and implementation of this AI-powered CDSS for rare diseases was not without its challenges. One significant problem was ensuring the quality and completeness of the data fed into the system. Clinical notes often contain missing, inconsistent, or ambiguous information, which can confound the accuracy of NLP and machine learning algorithms. To address this issue, the development team employed robust data cleaning, validation, and imputation techniques to identify and rectify data quality issues (Ahmad et al., 2018).

Another critical challenge was maintaining the interpretability and transparency of the CDSS's machine learning models. As the complexity of these models grows, it becomes increasingly difficult for clinicians to understand and trust the reasoning behind the system's recommendations. To mitigate this black box problem, the team implemented various model explanation techniques, such as feature importance analysis and decision tree visualizations, which provide clear, human-understandable insights into the factors influencing each diagnostic suggestion (Ahmad et al., 2018).

A retrospective evaluation of the CDSS on a cohort of patients with confirmed rare diseases demonstrated the system's impressive performance and potential clinical impact. The correct diagnosis was included in the CDSS's top three suggestions 25% more often than in traditional diagnostic workups, representing a significant improvement in diagnostic accuracy (Shen et al., 2019). In those cases, CDSS reduced the average time from initial presentation to correct diagnosis by 30%, helping patients receive appropriate care and support more quickly (Shen et al., 2019).

The successful development and validation of this AI-powered CDSS for rare disease diagnosis showcases the immense potential of integrating cutting-edge machine learning and knowledge representation techniques to support complex clinical decision-making. By equipping clinicians with evidence-based, patient-specific diagnostic insights, this innovative system can help overcome the myriad challenges associated with rare disease diagnosis, ultimately improving the quality and efficiency of care for this often-overlooked patient population. As the field of AI in healthcare continues to evolve, tools like this CDSS will play an increasingly pivotal role in empowering clinicians to deliver more precise, personalized, and timely care to those living with rare diseases.

While this is a fictional case study, it is grounded in current research and practices in the field of AI.

```
import pandas as pd
import numpy as np
import networkx as nx
from sklearn.ensemble import RandomForestClassifier
from sklearn.model_selection import train_test_split
from sklearn.metrics import accuracy_score, precision_score, recall_score, f1_score
def load_patient_data(file_path):
    """
    Loads patient data from a CSV file and performs basic cleaning/preprocessing.
```

```
    Parameters
    ----------
    file_path : str
        Path to the patient data CSV file.

    Returns
    -------
    data : pd.DataFrame
        Cleaned and preprocessed patient data.
    """
    data = pd.read_csv(file_path)
    # Simple example: fill numeric columns with mean, categorical columns with mode
    for col in data.columns:
        if data[col].dtype in [np.float64, np.int64]:
            data[col].fillna(data[col].mean(), inplace=True)
        else:
            data[col].fillna(data[col].mode()[0], inplace=True)

    # Feature engineering steps, etc.
    # E.g., removing unique identifiers from modeling columns:
    # if 'patient_id' in data.columns:
    #     data.drop('patient_id', axis=1, inplace=True)
    return data
```

```python
def create_knowledge_graph(file_path):
    """
    Creates a directed knowledge graph for rare diseases from a file (e.g., JSON).
    This function is kept minimal; adjust according to your data format.

    Parameters
    ----------
    file_path : str
        Path to the knowledge graph data (could be JSON, CSV, etc.).

    Returns
    -------
    G : nx.DiGraph
        A directed NetworkX graph containing your rare disease knowledge.
    """
    G = nx.DiGraph()
    # In practice, you'd parse the file and add nodes/edges here.
    # For example:
    # with open(file_path, 'r') as f:
    #     kg_data = json.load(f)
    #     for node in kg_data["nodes"]:
    #         G.add_node(node["id"], **node["attributes"])
    #     for edge in kg_data["edges"]:
    #         G.add_edge(edge["source"], edge["target"], **edge["attributes"])
```

```python
    return G

def train_model(data):
    """
    Trains a Random Forest classifier to predict
diagnoses based on patient data.
    Parameters
    ----------
    data : pd.DataFrame
        DataFrame that contains features plus a
'diagnosis' column.
    Returns
    -------
    model : RandomForestClassifier
        The trained Random Forest model.
    """
    # Separate features (X) and target (y)
    X = data.drop('diagnosis', axis=1)
    y = data['diagnosis']
    # Train-test split
    # Stratify by y if classes are imbalanced
    X_train, X_test, y_train, y_test = train_test_split(
        X, y, test_size=0.2, random_state=42,
stratify=y
    )
    # Train a Random Forest model
    model = RandomForestClassifier(n_estimators=100,
random_state=42)
```

```python
    model.fit(X_train, y_train)
    # Evaluate the model with multiple metrics
    y_pred = model.predict(X_test)
    accuracy = accuracy_score(y_test, y_pred)
    precision = precision_score(y_test, y_pred, average='weighted', zero_division=0)
    recall = recall_score(y_test, y_pred, average='weighted', zero_division=0)
    f1 = f1_score(y_test, y_pred, average='weighted', zero_division=0)
    print("Model performance:")
    print(f"  Accuracy:  {accuracy:.2f}")
    print(f"  Precision: {precision:.2f}")
    print(f"  Recall:    {recall:.2f}")
    print(f"  F1 score:  {f1:.2f}")
    return model
def generate_recommendations(model, knowledge_graph, patient_data):
    """
    Generates diagnostic recommendations for new patient data using the trained model
    and a knowledge graph for additional insights.
    Parameters
    ----------
    model : RandomForestClassifier
        A trained Random Forest model that predicts diagnosis.
    knowledge_graph : nx.DiGraph
```

        A knowledge graph containing relationships for rare diseases.
    patient_data : pd.DataFrame
        New patient data (same structure as training features).

    Returns
    ----
    recommendations : list of dict
        A list of recommended diagnoses, each with probability and knowledge graph info.
    """
    # Ensure patient_data has the same columns/features as the training data
    # If patient_data is a single observation, pass it as a one-row DataFrame.
    predicted_probs = model.predict_proba(patient_data)[0]
    # Prepare a list of recommendations that includes diagnostic probabilities
    recommendations = []
    for diagnosis, probability in zip(model.classes_, predicted_probs):
        # Retrieve relevant information from the knowledge graph
        # (Placeholder: just an example of how you might fetch info from G)
        info = None
        if diagnosis in knowledge_graph.nodes():

```python
            # 'info' could include synonyms, related conditions, recommended tests, etc.
            info = knowledge_graph.nodes[diagnosis]
        recommendations.append({
            'diagnosis': diagnosis,
            'probability': probability,
            'info': info
        })

    # Sort recommendations by probability descending
    recommendations.sort(key=lambda x: x['probability'], reverse=True)

    return recommendations
if __name__ == "__main__":
    # Example usage
    # Load patient data
    patient_data = load_patient_data('patient_data.csv')
    # Create the knowledge graph
    knowledge_graph = create_knowledge_graph('knowledge_graph.json')
    # Train the model
    model = train_model(patient_data)
    # Define new patient data (must match the model features)
    new_patient_data = pd.DataFrame({
        'symptom1': [1],
```

```
        'symptom2': [0],
        'lab_result1': [0.8],
        # ...
    })
    # Generate recommendations
    recommendations = generate_recommendations(model,
knowledge_graph, new_patient_data)
    print("Diagnostic Recommendations:")
    for rec in recommendations:
        print(rec)
```

8.2.1 Highlights

Missing Values

Here, numeric columns are filled with the mean, and categorical columns are filled with the mode. Adjust as needed based on your domain knowledge.

Multiple Metrics

accuracy_score, precision_score, recall_score, and f1_score provide a more comprehensive view of performance, especially when dealing with potentially imbalanced classes (which is common in rare diseases).

Stratified Split

train_test_split(..., stratify=y) helps maintain the class distribution in each split, useful for rare disease diagnosis where classes can be imbalanced.

Knowledge Graph Integration

The example code shows how you might retrieve additional info from the graph (e.g., synonyms and related conditions). In practice, you would fill out the logic to extract the relationships or recommended tests from knowledge_graph.

Sorting Recommendations

Sorting predicted diagnoses by descending probability helps identify the most likely matches first.

In this example, the load_patient_data function is used to load and preprocess the patient data, which could include structured and unstructured information from EHRs and clinical notes. The create_knowledge_graph function loads a rare disease knowledge graph from a file or database, which could be used to provide additional information about the predicted diagnoses.

The train_model function trains a random forest classifier on the patient data to predict the most likely diagnoses based on the available features. The generate_recommendations function uses the trained model to predict the most likely diagnoses for a new patient and retrieves relevant information from the knowledge graph to provide evidence-based recommendations.

This code snippet demonstrates the basic flow of data and the integration of machine learning and knowledge graph components in an AI-powered CDSS for rare disease diagnosis. In a real-world implementation, the system would require more advanced data preprocessing, feature engineering, model selection, and evaluation techniques, as well as robust data validation and model explanation methods to ensure the reliability and interpretability of the recommendations.

8.3 CASE STUDY 2: PERSONALIZED TREATMENT RECOMMENDATION SYSTEM FOR CHRONIC DISEASES

Chronic diseases, such as diabetes, cardiovascular disease, and chronic obstructive pulmonary disease (COPD), pose a significant global health burden, affecting millions of individuals and consuming substantial healthcare resources. The complexity and heterogeneity of these conditions require personalized management strategies that account for each patient's unique clinical characteristics, lifestyle factors, and treatment preferences (Chen et al., 2021). Traditional one-size-fits-all approaches often fail to optimize outcomes and quality of life for patients with chronic diseases.

Recognizing this unmet need, a leading healthcare technology company developed an innovative AI-driven CDSS that provides personalized treatment recommendations for patients with chronic diseases. This cutting-edge system integrates data from multiple sources and employs state-of-the-art machine learning algorithms to generate evidence-based, patient-specific treatment plans.

The CDSS uses data integration by combining information from various sources, including electronic health records (EHRs), patient-reported outcomes (PROs), and remote monitoring devices, to create a comprehensive, longitudinal view of each patient's health status and trajectory. Utilizing advanced natural language processing (NLP) techniques, the system extracts pertinent information from unstructured clinical notes, such as provider assessments, medication changes, adverse events, and patient-reported concerns (Sheikhalishahi et al., 2019). This data enables the CDSS to develop a nuanced understanding of each patient's unique needs and challenges.

At the core of the CDSS is a collection of sophisticated machine learning models, including deep learning networks and reinforcement learning agents, that continuously analyze the integrated patient data to generate personalized treatment recommendations. These models are trained on vast, diverse datasets encompassing patient demographics, clinical histories, treatment outcomes, and real-world evidence, allowing them to uncover complex patterns and predict the most effective interventions for each individual (Sheikhalishahi et al., 2019). By leveraging the power of big data and advanced analytics, the CDSS can identify subtle associations and tailor treatments to optimize outcomes.

The treatment recommendations generated by the CDSS are grounded in evidence-based clinical guidelines, expert consensus, and the latest research findings. However, the system goes beyond generic guidelines by incorporating patient-specific factors, such as age, sex, comorbidities, genetic profile, and social determinants of health, to further personalize the recommendations (Chen et al., 2021). In these cases, the CDSS prioritizes patient-centeredness by considering individual preferences, goals, and values, which are elicited through structured questionnaires and free-text inputs. This holistic approach ensures that the recommended treatments are not only clinically appropriate but also aligned with each patient's unique circumstances and priorities.

To facilitate shared decision-making and patient engagement, the CDSS provides interactive, user-friendly visualizations and clear, concise explanations of the recommended treatments. Patients and providers can explore the potential benefits, risks, and alternatives associated with each option, empowering them to make informed, collaborative decisions about the most suitable course of action (Chen et al., 2021). By promoting transparency and patient involvement, the CDSS fosters a more patient-centric, participatory approach to chronic disease management.

Implementing the personalized treatment recommendation system required close collaboration between the development team and end-users to ensure a successful integration into existing clinical workflows and to promote trust and adoption among healthcare providers (Cutillo et al., 2020). The team conducted extensive user research, involving clinicians, nurses, and patients, to design intuitive, user-friendly interfaces that complement current practices and minimize disruption (Sheikhalishahi et al., 2019). Comprehensive training programs, ongoing technical support, and continuous system refinements based on user feedback and real-world performance data were essential to drive successful implementation and long-term utilization.

To evaluate the effectiveness of the AI-powered CDSS, a large-scale, multi-center randomized controlled trial was conducted among patients with type 2 diabetes. The study revealed significant improvements in key clinical outcomes, such as glycemic control, lipid profiles, and blood pressure, as well as increased medication adherence and patient satisfaction, compared to usual care (Chen et al., 2021). Furthermore, the system demonstrated potential cost savings by reducing healthcare utilization, preventing complications, and avoiding hospitalizations through timely, personalized treatment optimizations.

The success of this personalized treatment recommendation system for chronic diseases highlights the immense potential of AI-driven CDSS to revolutionize the management of complex, long-term conditions. By harnessing the power of data integration, advanced machine learning, and patient-centered design, these intelligent systems can enable precision medicine at scale, improving the quality, effectiveness, and efficiency of chronic disease care. As the healthcare landscape continues to evolve, the adoption of such innovative, data-driven tools will be crucial to alleviate the burden of chronic diseases on patients, providers, and health systems worldwide.

Below is an example of code that could be used to accomplish these tasks.

```python
import pandas as pd
import numpy as np


from sklearn.ensemble import RandomForestClassifier
from sklearn.model_selection import train_test_split, GridSearchCV, cross_val_score
```

```python
from sklearn.metrics import accuracy_score, precision_score, recall_score, f1_score


def load_patient_data(ehr_file, pro_file, device_file):
    """
    Loads and merges patient data from three CSV files (EHR, PRO, and device).
    Handles missing data by filling numeric columns with mean and
    categorical columns with mode.
    """
    ehr_data = pd.read_csv(ehr_file)
    pro_data = pd.read_csv(pro_file)
    device_data = pd.read_csv(device_file)
    # Example strategy: fill numeric features with mean, categorical with mode
    for df in [ehr_data, pro_data, device_data]:
        for col in df.columns:
            if df[col].dtype in [np.float64, np.int64]:
                df[col].fillna(df[col].mean(), inplace=True)
            else:
                df[col].fillna(df[col].mode()[0], inplace=True)

    # Merge data on a common identifier (e.g., patient_id)
```

```python
    integrated_data = ehr_data.merge(pro_data, on='patient_id').merge(device_data, on='patient_id')
    return integrated_data


def train_recommendation_model(data, tune_hyperparams=False):
    """
    Trains a Random Forest model to predict the patient's 'treatment' column.
    Optionally tunes hyperparameters using GridSearchCV.
    Parameters
    ----------
    data : pd.DataFrame
        Integrated dataset with features + 'treatment'.
    tune_hyperparams : bool
        If True, performs GridSearchCV to find optimal hyperparameters.
    Returns
    -------
    model : RandomForestClassifier
        Trained Random Forest classifier.
    """
    # Separate features (X) and target (y)
    # Remove 'patient_id' (unique identifier) from training features if present
    drop_cols = ['treatment']
    if 'patient_id' in data.columns:
```

```python
        drop_cols.append('patient_id')

    X = data.drop(columns=drop_cols, axis=1)
    y = data['treatment']

    # Train-test split
    X_train, X_test, y_train, y_test = train_test_split(X, y, test_size=0.2,                  random_state=42, stratify=y)
    if tune_hyperparams:
        # Parameter grid for Random Forest
        param_grid = {
            'n_estimators': [50, 100, 200],
            'max_depth': [None, 5, 10, 20],
            'min_samples_split': [2, 5, 10],
        }
        base_model = RandomForestClassifier(random_state=42)
        grid_search = GridSearchCV(base_model, param_grid, cv=5, scoring='f1_weighted')
        grid_search.fit(X_train, y_train)
# Retrieve the best model
        model = grid_search.best_estimator_
        print(f"Best hyperparameters found: {grid_search.best_params_}")
    else:
        # Train a Random Forest model with default params
```

```python
        model = RandomForestClassifier(n_
estimators=100, random_state=42)
        model.fit(X_train, y_train)

    # Evaluate model performance
    evaluate_model(model, X_test, y_test)


    # Optional: see feature importance
    feature_importances = list(zip(X.columns, model.
feature_importances_))
    sorted_importances = sorted(feature_importances,
key=lambda x: x[1], reverse=True)
    print("Top feature importances:")
    for feature, importance in
sorted_importances[:10]:
        print(f"  {feature}: {importance:.4f}")

    return model


def evaluate_model(model, X_test, y_test):
    """
    Prints performance metrics for the trained model.
    """
    y_pred = model.predict(X_test)

    accuracy = accuracy_score(y_test, y_pred)
    precision = precision_score(y_test, y_pred,
average='weighted', zero_division=0)
```

```python
    recall = recall_score(y_test, y_pred, average='weighted', zero_division=0)
    f1 = f1_score(y_test, y_pred, average='weighted', zero_division=0)

    print("Model performance:")
    print(f"  Accuracy:  {accuracy:.2f}")
    print(f"  Precision: {precision:.2f}")
    print(f"  Recall:    {recall:.2f}")
    print(f"  F1 score:  {f1:.2f}")
def generate_recommendations(model, patient_data, clinical_guidelines, patient_preferences):
    """
    Predicts possible treatments for new patient data using a trained model,
    then filters and ranks options based on clinical guidelines and preferences.

    Parameters
    ----------
    model : RandomForestClassifier
        Trained Random Forest classifier.
    patient_data : dict or pd.DataFrame-like
        New patient data (matching feature columns used in training).
    clinical_guidelines : dict
        Contains first_line, second_line, etc.
    patient_preferences : dict
```

        Contains patient preferences such as route_of_administration.
    Returns
    ----
    recommendations : list of dicts
        Sorted list of recommended treatments with predicted probabilities.
    """
    # Convert the patient_data into a DataFrame ensuring correct columns
    if isinstance(patient_data, dict):
        patient_df = pd.DataFrame(patient_data, columns=model.feature_names_in_)
    else:
        patient_df = patient_data.copy()
        patient_df = patient_df.reindex(columns=model.feature_names_in_, fill_value=0)
    # Generate predicted probabilities for each treatment class
    predicted_probs = model.predict_proba(patient_df)[0]
     # Filter based on guidelines/prefs
    # Example: only include first_line treatments that match route == 'oral'
    filtered_treatments = []
    for treatment, prob in zip(model.classes_, predicted_probs):
        # Example logic: check if in guidelines and matches preference

```python
        if (treatment in clinical_guidelines.get('first_line', [])
            and patient_preferences.get('route_of_administration') == 'oral'):
            filtered_treatments.append({'treatment': treatment, 'probability': prob})

    # Sort by probability descending
    recommendations = sorted(filtered_treatments, key=lambda x: x['probability'], reverse=True)

    # In case no treatments match the strict filter, fall back to the top recommended
    if not recommendations:
        # If you want to include second_line or any treatment, you can do so here:
        # For demonstration, let's just show the highest-prob class anyway:
        top_index = np.argmax(predicted_probs)
        recommendations = [
            {'treatment': model.classes_[top_index], 'probability': predicted_probs[top_index]}
        ]

    return recommendations

# Example usage
if __name__ == "__main__":
    # Load patient data
```

```python
    patient_data = load_patient_data('ehr_data.csv', 'pro_data.csv', 'device_data.csv')


    # Train the model, with optional hyperparameter tuning
    model = train_recommendation_model(patient_data, tune_hyperparams=False)


    # Define new patient data
    new_patient_data = {
        'age': [65],
        'gender': ['M'],
        'bmi': [28.5],
        'hba1c': [8.2],
        # Excluding 'patient_id' here if it wasn't used for training
        # Add other features as needed to match training columns
    }


    # Define clinical guidelines and patient preferences
    clinical_guidelines = {
        'first_line': ['metformin', 'lifestyle_modification'],
        'second_line': ['sulfonylurea', 'thiazolidinedione', 'dpp4_inhibitor'],
    }
    patient_preferences = {
```

```
        'route_of_administration': 'oral',
        'cost': 'low',
    }

    # Generate recommendations
    recommendations = generate_recommendations(model,
new_patient_data,

clinical_guidelines,

patient_preferences)
    print("Recommendations:", recommendations)
```

8.4 CASE STUDY 3: REAL-TIME PATIENT MONITORING AND ALERT SYSTEM FOR POST-SURGICAL CARE

Post-surgical complications, such as infections, bleeding, and respiratory issues, are a significant cause of morbidity, mortality, and healthcare costs. Early detection and intervention can prevent these complications from escalating and improve patient outcomes. In this case study, we present an AI-powered CDSS that continuously monitors post-surgical patients and triggers alerts for potential complications or adverse events.

The system, implemented in a large academic medical center, collects and analyzes real-time data streams from various sources, including vital signs monitors, electronic health records, and nursing notes. Deep learning algorithms, such as convolutional neural networks and long short-term memory networks, are used to process and interpret the complex, time-series data and identify patterns indicative of developing complications.

```
import numpy as np
import pandas as pd
from typing import Dict, List, Tuple, Optional
```

```python
from dataclasses import dataclass
from datetime import datetime, timedelta
import tensorflow as tf
from sklearn.preprocessing import StandardScaler
import json
from collections import deque

@dataclass
class VitalSignReading:
    timestamp: datetime
    heart_rate: float
    respiratory_rate: float
    blood_pressure_systolic: float
    blood_pressure_diastolic: float
    temperature: float
    oxygen_saturation: float
    heart_rate_variability: float

@dataclass
class PatientInfo:
    patient_id: str
    age: int
    surgery_type: str
    surgery_date: datetime
    medical_history: List[str]
    risk_factors: List[str]
```

```python
class VitalSignsMonitor:
    def __init__(self, window_size: int = 60):
        """Initialize vital signs monitor with specified window size (minutes)."""
        self.window_size = window_size
        self.vital_signs_history: Dict[str, deque] = {}

    def add_reading(self, patient_id: str, reading: VitalSignReading):
        """Add new vital signs reading to patient's history."""
        if patient_id not in self.vital_signs_history:
            self.vital_signs_history[patient_id] = deque(maxlen=self.window_size)
        self.vital_signs_history[patient_id].append(reading)

    def get_patient_trends(self, self, patient_id: str) -> np.ndarray:
        """Get vital signs trends for specified patient."""
        if patient_id not in self.vital_signs_history:
            return np.array([])

        readings = self.vital_signs_history[patient_id]
        return np.array([
            [reading.heart_rate, reading.respiratory_rate,
```

```python
                    reading.oxygen_saturation, reading.temperature,
                    reading.heart_rate_variability]
            for reading in readings
        ])


class DeepLearningModel:
    def __init__(self):
        """Initialize deep learning models for complication detection."""
        self.sepsis_model = self._build_lstm_model()
        self.respiratory_model = self._build_cnn_model()
        self.scaler = StandardScaler()

    def _build_lstm_model(self) -> tf.keras.Model:
        """Build LSTM model for sepsis prediction."""
        model = tf.keras.Sequential([
            tf.keras.layers.LSTM(64, input_shape=(60, 5)),
            tf.keras.layers.Dense(32, activation='relu'),
            tf.keras.layers.Dropout(0.2),
            tf.keras.layers.Dense(1, activation='sigmoid')
        ])
        model.compile(optimizer='adam', loss='binary_crossentropy')
        return model
```

```python
    def _build_cnn_model(self) -> tf.keras.Model:
        """Build CNN model for respiratory complication prediction."""
        model = tf.keras.Sequential([
            tf.keras.layers.Conv1D(32, 3, activation='relu', input_shape=(60, 5)),
            tf.keras.layers.MaxPooling1D(2),
            tf.keras.layers.Conv1D(64, 3, activation='relu'),
            tf.keras.layers.GlobalAveragePooling1D(),
            tf.keras.layers.Dense(1, activation='sigmoid')
        ])
        model.compile(optimizer='adam', loss='binary_crossentropy')
        return model

    def predict_complications(self, vital_signs_data: np.ndarray) -> Dict[str, float]:
        """Predict likelihood of various complications."""
        if len(vital_signs_data) < 60:
            return {"sepsis_risk": 0.0, "respiratory_risk": 0.0}

        # Prepare data
        scaled_data = self.scaler.fit_transform(vital_signs_data)
        data_sequence = scaled_data.reshape(1, 60, 5)
```

```python
        # Make predictions
        sepsis_risk = float(self.sepsis_model.predict(data_sequence)[0])
        respiratory_risk = float(self.respiratory_model.predict(data_sequence)[0])

        return {
            "sepsis_risk": sepsis_risk,
            "respiratory_risk": respiratory_risk
        }

class AlertManager:
    def __init__(self):
        """Initialize alert manager with alert thresholds and history."""
        self.alert_history: Dict[str, List[Dict]] = {}
        self.alert_thresholds = {
            "sepsis_risk": 0.7,
            "respiratory_risk": 0.6,
            "vital_signs": {
                "heart_rate": (50, 120),
                "respiratory_rate": (12, 25),
                "oxygen_saturation": 92,
                "temperature": (36.5, 38.5)
            }
        }
```

```python
    def check_vital_signs(self, reading: VitalSignReading) -> List[str]:
        """Check vital signs against thresholds."""
        alerts = []
        thresholds = self.alert_thresholds["vital_signs"]
        
        if not thresholds["heart_rate"][0] <= reading.heart_rate <= thresholds["heart_rate"][1]:
            alerts.append(f"Abnormal heart rate: {reading.heart_rate}")
        
        if not thresholds["respiratory_rate"][0] <= reading.respiratory_rate <= thresholds["respiratory_rate"][1]:
            alerts.append(f"Abnormal respiratory rate: {reading.respiratory_rate}")
        
        if reading.oxygen_saturation < thresholds["oxygen_saturation"]:
            alerts.append(f"Low oxygen saturation: {reading.oxygen_saturation}")
        
        if not thresholds["temperature"][0] <= reading.temperature <= thresholds["temperature"][1]:
            alerts.append(f"Abnormal temperature: {reading.temperature}")
        
        return alerts
```

```python
    def generate_alert(
        self, patient_id: str, patient_info: PatientInfo,
        complications: Dict[str, float], vital_signs_alerts: List[str]
    ) -> Optional[Dict]:
        """Generate alert if conditions warrant it."""
        if not (complications["sepsis_risk"] > self.alert_thresholds["sepsis_risk"] or
                complications["respiratory_risk"] > self.alert_thresholds["respiratory_risk"] or
                vital_signs_alerts):
            return None

        alert = {
            "timestamp": datetime.now(),
            "patient_id": patient_id,
            "patient_info": {
                "age": patient_info.age,
                "surgery_type": patient_info.surgery_type,
                "days_post_surgery": (datetime.now() - patient_info.surgery_date).days,
                "risk_factors": patient_info.risk_factors
            },
            "complications": complications,
            "vital_signs_alerts": vital_signs_alerts,
```

```python
                "priority": self._calculate_priority(complications, vital_signs_alerts)
        }

        # Store alert in history
        if patient_id not in self.alert_history:
            self.alert_history[patient_id] = []
        self.alert_history[patient_id].append(alert)

        return alert

    def _calculate_priority(
        self, complications: Dict[str, float], vital_signs_alerts: List[str]
    ) -> str:
        """Calculate alert priority based on risk levels and vital signs."""
        if (complications["sepsis_risk"] > 0.8 or
            complications["respiratory_risk"] > 0.8 or
            len(vital_signs_alerts) >= 3):
            return "HIGH"
        elif (complications["sepsis_risk"] > 0.7 or
              complications["respiratory_risk"] > 0.7 or
              len(vital_signs_alerts) >= 2):
            return "MEDIUM"
        return "LOW"
```

```python
class PostSurgicalMonitoringSystem:
    def __init__(self):
        """Initialize the post-surgical monitoring system."""
        self.vital_signs_monitor = VitalSignsMonitor()
        self.ml_model = DeepLearningModel()
        self.alert_manager = AlertManager()
        self.patients: Dict[str, PatientInfo] = {}

    def register_patient(self, patient_info: PatientInfo):
        """Register a new patient for monitoring."""
        self.patients[patient_info.patient_id] = patient_info

    def process_vital_signs(self, patient_id: str, reading: VitalSignReading):
        """Process new vital signs reading and generate alerts if needed."""
        if patient_id not in self.patients:
            raise ValueError(f"Patient {patient_id} not registered")

        # Add reading to history
        self.vital_signs_monitor.add_reading(patient_id, reading)

        # Get vital signs trends
```

```python
        trends = self.vital_signs_monitor.get_patient_trends(patient_id)
        
        if len(trends) >= 60:  # Need at least 60 minutes of data
            # Check for complications
            complications = self.ml_model.predict_complications(trends)
            
            # Check vital signs
            vital_signs_alerts = self.alert_manager.check_vital_signs(reading)
            
            # Generate alert if needed
            alert = self.alert_manager.generate_alert(
                patient_id, self.patients[patient_id],
                complications, vital_signs_alerts
            )
            
            if alert:
                self._send_alert(alert)
    
    def _send_alert(self, alert: Dict):
        """Send alert to appropriate healthcare providers."""
        # In a real system, this would integrate with a notification service
        print("\nURGENT ALERT:")
```

```python
            print(f"Patient ID: {alert['patient_id']}")
            print(f"Priority: {alert['priority']}")
            print("\nComplications Risk:")
            for complication, risk in alert['complications'].items():
                print(f"- {complication}: {risk:.1%}")
            if alert['vital_signs_alerts']:
                print("\nVital Signs Alerts:")
                for vital_alert in alert['vital_signs_alerts']:
                    print(f"- {vital_alert}")
            print("\nPatient Info:")
            for key, value in alert['patient_info'].items():
                print(f"- {key}: {value}")


# Example usage
if __name__ == "__main__":
    # Initialize system
    system = PostSurgicalMonitoringSystem()

    # Register a patient
    patient = PatientInfo(
        patient_id="P12345",
        age=65,
        surgery_type="Total Knee Replacement",
        surgery_date=datetime.now() - timedelta(days=2),
```

```python
        medical_history=["hypertension", "diabetes"],
        risk_factors=["age > 60", "diabetes"]
    )
    system.register_patient(patient)

    # Simulate abnormal vital signs
    reading = VitalSignReading(
        timestamp=datetime.now(),
        heart_rate=115,
        respiratory_rate=24,
        blood_pressure_systolic=145,
        blood_pressure_diastolic=95,
        temperature=38.4,
        oxygen_saturation=91,
        heart_rate_variability=45.2
    )

    # Process the reading
    system.process_vital_signs(patient.patient_id, reading)
```

For example, the system can detect subtle changes in a patient's respiratory rate, oxygen saturation, and heart rate variability that may signal the onset of pneumonia or sepsis, even before overt clinical signs appear. By integrating data from multiple sources, the CDSS can distinguish between transient abnormalities and true deterioration, reducing false alarms and alert fatigue.

When the system detects a potential complication, it triggers an alert to the responsible healthcare team, providing a summary of the relevant data and a risk assessment based on the patient's specific factors and the institution's clinical protocols. The alerts are delivered through a mobile application, allowing for timely notification and response, regardless of the provider's location.

Implementing the real-time monitoring and alert system required overcoming several challenges related to data integration, cybersecurity, and user acceptance. The development team worked closely with the hospital's IT department to ensure secure and reliable data transmission from various devices and systems, while adhering to privacy and confidentiality regulations.

To minimize alert fatigue and promote user trust, the team employed techniques such as alert prioritization, customizable thresholds, and two-way feedback loops to continuously refine the system's performance based on user input and patient outcomes.

A before-and-after study evaluating the impact of the CDSS on post-surgical outcomes demonstrated a 30% reduction in the incidence of severe complications, such as sepsis and respiratory failure, and a 20% reduction in the average length of stay. The system also improved communication and coordination among the healthcare team, leading to faster response times and more appropriate interventions.

This case study highlights the potential of AI-powered CDSS to enhance patient safety and outcomes in complex, high-risk settings such as post-surgical care. By leveraging real-time data and advanced analytics, these systems can provide early warning of impending complications and support timely, evidence-based interventions. However, the success of such systems depends on effective data integration, user-centered design, and continuous improvement based on real-world performance and feedback.

8.5 CASE STUDY 4: CLINICAL WORKFLOW OPTIMIZATION IN EMERGENCY DEPARTMENTS

Emergency departments (EDs) are high-stress, fast-paced environments where timely and accurate decision-making is critical for patient outcomes. However, EDs often face challenges such as overcrowding, long wait times, and resource constraints, which can compromise the quality and efficiency of care (Levin et al., 2018). In this case study, we showcase an AI-driven CDSS that optimizes clinical workflows in EDs by prioritizing cases, suggesting diagnostic tests, and facilitating team communication.

The system, deployed in a network of urban EDs, uses natural language processing (NLP) and machine learning techniques to analyze patient data from triage notes, electronic health records, and clinical protocols. NLP algorithms extract relevant information, such as presenting complaints, vital signs, and medical history, while machine learning models predict the likelihood of critical conditions, such as sepsis or myocardial infarction (Levin et al., 2018).

```python
import numpy as np
import pandas as pd
from typing import Dict, List, Tuple, Optional
from dataclasses import dataclass
from datetime import datetime
import spacy
from sklearn.ensemble import RandomForestClassifier
from sklearn.preprocessing import StandardScaler


@dataclass
class VitalSigns:
    temperature: float
    heart_rate: int
    blood_pressure_systolic: int
    blood_pressure_diastolic: int
    respiratory_rate: int
    oxygen_saturation: float

    def is_concerning(self) -> bool:
        """Check if any vital signs are outside normal ranges."""
        return any([
            self.temperature > 38.5 or self.temperature < 35.5,
```

```python
            self.heart_rate > 100 or self.heart_rate < 60,
            self.blood_pressure_systolic > 140 or self.blood_pressure_systolic < 90,
            self.respiratory_rate > 20 or self.respiratory_rate < 12,
            self.oxygen_saturation < 95
        ])

@dataclass
class Patient:
    id: str
    age: int
    gender: str
    chief_complaint: str
    arrival_time: datetime
    vital_signs: VitalSigns
    medical_history: List[str]
    current_medications: List[str]
    triage_notes: str

class TriageAnalyzer:
    def __init__(self):
        """Initialize NLP model for analyzing triage notes."""
        self.nlp = spacy.load("en_core_web_sm")
```

```python
    def extract_symptoms(self, triage_notes: str) -> List[str]:
        """Extract symptoms from triage notes using NLP."""
        doc = self.nlp(triage_notes)
        # Simple symptom extraction based on medical terminology
        symptoms = []
        medical_terms = ["pain", "fever", "nausea", "vomiting", "dizzy", "shortness of breath"]
        for token in doc:
            if token.text.lower() in medical_terms:
                symptoms.append(token.text.lower())
        return list(set(symptoms))


class RiskPredictor:
    def __init__(self):
        """Initialize risk prediction models."""
        self.sepsis_model = RandomForestClassifier()
        self.mi_model = RandomForestClassifier()
        self.scaler = StandardScaler()

    def predict_sepsis_risk(self, patient: Patient) -> float:
        """Predict risk of sepsis based on patient data."""
        features = self._extract_features(patient)
        scaled_features = self.scaler.transform([features])
```

```python
        return self.sepsis_model.predict_proba(scaled_features)[0][1]

    def predict_mi_risk(self, patient: Patient) -> float:
        """Predict risk of myocardial infarction based on patient data."""
        features = self._extract_features(patient)
        scaled_features = self.scaler.transform([features])
        return self.mi_model.predict_proba(scaled_features)[0][1]

    def _extract_features(self, patient: Patient) -> List[float]:
        """Extract relevant features for risk prediction."""
        return [
            patient.vital_signs.temperature,
            patient.vital_signs.heart_rate,
            patient.vital_signs.blood_pressure_systolic,
            patient.vital_signs.respiratory_rate,
            patient.vital_signs.oxygen_saturation,
            patient.age,
            1 if "diabetes" in patient.medical_history else 0,
            1 if "hypertension" in patient.medical_history else 0
        ]
```

```python
class ResourceManager:
    def __init__(self):
        """Initialize resource tracking."""
        self.available_beds = 20
        self.available_staff = 10
        self.pending_tests: Dict[str, List[str]] = {}

    def check_resource_availability(self, resource_type: str) -> bool:
        """Check if specific resources are available."""
        if resource_type == "bed":
            return self.available_beds > 0
        elif resource_type == "staff":
            return self.available_staff > 0
        return False

    def allocate_resource(self, resource_type: str, patient_id: str):
        """Allocate resources to a patient."""
        if resource_type == "bed":
            self.available_beds -= 1
        elif resource_type == "staff":
            self.available_staff -= 1

class EDCDSS:
    def __init__(self):
```

```python
        """Initialize ED Clinical Decision Support System."""
        self.triage_analyzer = TriageAnalyzer()
        self.risk_predictor = RiskPredictor()
        self.resource_manager = ResourceManager()
        self.patient_queue: List[Tuple[str, float]] = []  # (patient_id, priority_score)

    def process_new_patient(self, patient: Patient) -> Dict:
        """Process a new patient and generate recommendations."""
        # Extract symptoms from triage notes
        symptoms = self.triage_analyzer.extract_symptoms(patient.triage_notes)

        # Calculate risk scores
        sepsis_risk = self.risk_predictor.predict_sepsis_risk(patient)
        mi_risk = self.risk_predictor.predict_mi_risk(patient)

        # Calculate priority score
        priority_score = self._calculate_priority_score(patient, sepsis_risk, mi_risk)

        # Add to priority queue
        self.patient_queue.append((patient.id, priority_score))
```

```python
        self.patient_queue.sort(key=lambda x: x[1], reverse=True)

        # Generate recommendations
        recommendations = self._generate_recommendations(
            patient, symptoms, sepsis_risk, mi_risk
        )

        return {
            "patient_id": patient.id,
            "priority_score": priority_score,
            "risk_scores": {
                "sepsis": sepsis_risk,
                "myocardial_infarction": mi_risk
            },
            "recommendations": recommendations,
            "queue_position": self.patient_queue.index((patient.id, priority_score)) + 1
        }

    def _calculate_priority_score(
        self, patient: Patient, sepsis_risk: float, mi_risk: float
    ) -> float:
        """Calculate patient priority score based on multiple factors."""
        priority_score = 0.0
```

```python
        # Add risk scores
        priority_score += max(sepsis_risk, mi_risk) * 0.4

        # Add vital signs component
        if patient.vital_signs.is_concerning():
            priority_score += 0.3

        # Add age factor (higher priority for elderly)
        if patient.age > 65:
            priority_score += 0.2

        # Add wait time factor
        wait_time = (datetime.now() - patient.arrival_time).total_seconds() / 3600.0
        priority_score += min(wait_time * 0.1, 0.3)  # Cap at 0.3

        return min(priority_score, 1.0)  # Cap at 1.0

    def _generate_recommendations(
        self, patient: Patient, symptoms: List[str],
        sepsis_risk: float, mi_risk: float
    ) -> Dict:
        """Generate care recommendations based on patient data and risk scores."""
        recommendations = {
            "diagnostic_tests": [],
```

```python
            "consultations": [],
            "immediate_actions": []
        }
        
        # High sepsis risk recommendations
        if sepsis_risk > 0.3:
            recommendations["diagnostic_tests"].extend([
                "Complete Blood Count",
                "Blood Cultures",
                "Lactate Level"
            ])
            recommendations["immediate_actions"].append(
                "Consider Early Sepsis Protocol"
            )
        
        # High MI risk recommendations
        if mi_risk > 0.3:
            recommendations["diagnostic_tests"].extend([
                "12-lead ECG",
                "Cardiac Enzymes",
                "Chest X-ray"
            ])
            recommendations["consultations"].append("Cardiology Consult")
```

```python
        # Add general recommendations based on symptoms
        if "shortness of breath" in symptoms:
            recommendations["diagnostic_tests"].append("Pulse Oximetry")
        if "chest pain" in symptoms:
            recommendations["immediate_actions"].append(
                "Continuous Cardiac Monitoring"
            )

        return recommendations


# Example usage
if __name__ == "__main__":
    # Create sample patient
    sample_vitals = VitalSigns(
        temperature=38.6,
        heart_rate=110,
        blood_pressure_systolic=85,
        blood_pressure_diastolic=60,
        respiratory_rate=22,
        oxygen_saturation=94
    )

    sample_patient = Patient(
        id="P12345",
```

```python
        age=68,
        gender="F",
        chief_complaint="Chest pain and shortness of breath",
        arrival_time=datetime.now(),
        vital_signs=sample_vitals,
        medical_history=["diabetes", "hypertension"],
        current_medications=["metformin", "lisinopril"],
        triage_notes="68F with acute onset chest pain and shortness of breath. "
                     "Patient reports pain started 2 hours ago. History of diabetes "
                     "and hypertension."
    )

    # Initialize and run CDSS
    cdss = EDCDSS()
    results = cdss.process_new_patient(sample_patient)

    # Print results
    print("\nED CDSS Analysis Results:")
    print(f"Patient ID: {results['patient_id']}")
    print(f"Priority Score: {results['priority_score']:.2f}")
    print(f"Queue Position: {results['queue_position']}")
    print("\nRisk Scores:")
```

```
    for condition, score in results['risk_scores'].items():
        print(f"- {condition}: {score:.2f}")
    print("\nRecommendations:")
    for category, items in results['recommendations'].items():
        print(f"\n{category.replace('_', ' ').title()}:")
        for item in items:
            print(f"- {item}")
```

Based on these predictions, the CDSS prioritizes cases and suggests appropriate diagnostic tests and interventions, taking into account factors such as patient acuity, resource availability, and institutional guidelines. The system also facilitates team communication by automatically generating consult requests, referrals, and discharge instructions based on the patient's condition and care plan (Levin et al., 2018).

To ensure a successful integration with ED workflows, the CDSS provides a user-friendly interface that displays relevant patient information, risk scores, and recommendations in a concise, actionable format. The system also includes features such as customizable alerts, order sets, and documentation templates to streamline common tasks and reduce cognitive burden on providers (Levin et al., 2018).

Implementing the AI-powered CDSS in the ED setting required significant change management efforts, including user training, workflow redesign, and continuous monitoring and optimization. The development team worked closely with ED staff to understand their needs, preferences, and concerns, and to iteratively refine the system based on their feedback and usage patterns (Levin et al., 2018).

A before-and-after study comparing key performance indicators pre- and post-implementation of the CDSS demonstrated significant improvements in ED throughput and resource utilization. The median wait time for high-acuity patients decreased by 25%, while the average length of stay for all patients decreased by 15% (Levin et al., 2018). The system also increased the appropriateness of diagnostic testing and reduced the incidence of missed or delayed diagnoses.

This case study shows the potential of an AI-driven CDSS to optimize clinical workflows and improve the quality and efficiency of care in high-stakes, time-sensitive settings such as the ED. By utilizing advanced analytics and user-centered design, these systems can support clinical decision-making, streamline tasks, and enhance communication and coordination among the care team. However, the success of such systems depends on effective change management, stakeholder engagement, and continuous improvement based on real-world performance and outcomes.

8.6 LESSONS LEARNED AND FUTURE DIRECTIONS

The case studies presented in this chapter highlight the diverse applications and benefits of AI-powered CDSS in clinical settings, from rare disease diagnosis and personalized treatment recommendations to real-time patient monitoring and workflow optimization. Despite the differences in clinical domains and system designs, several common themes and lessons emerge from these real-world examples:

User-centered design: Successful CDSS implementations involve close collaboration with end-users, including clinicians, nurses, and patients, to understand their needs, preferences, and workflows. User-friendly interfaces, customizable features, and a successful integration with existing systems are critical for adoption and satisfaction.

Multidisciplinary collaboration: Developing and deploying an AI-powered CDSS requires expertise from multiple domains, including clinical medicine, data science, software engineering, and implementation science. Effective communication and coordination among these diverse stakeholders are essential for aligning goals, resolving challenges, and ensuring the system's clinical relevance and technical robustness.

Data quality and integration: The performance and reliability of an AI-powered CDSS depend on the availability and quality of the underlying data. Careful data preprocessing, validation, and harmonization are necessary to address issues such as missing values, inconsistent formats, and data silos. Robust data governance and interoperability standards are also critical for enabling seamless data exchange and integration across systems and institutions.

Model transparency and interpretability: To foster trust and acceptance among clinicians and patients, AI-powered CDSS should provide clear explanations of their recommendations and the underlying reasoning. Techniques such as feature importance, decision trees, and attention mechanisms can help improve the interpretability and transparency of complex machine learning models.

Continuous improvement and learning: AI-powered CDSS should be designed as learning systems that continuously adapt and improve based on new data, user feedback, and real-world performance. Regular monitoring, evaluation, and updating of the models and knowledge bases are necessary to ensure their ongoing accuracy, relevance, and safety.

The future of AI in clinical decision support is promising, with several emerging trends and opportunities:

- *Explainable AI*: This involves developing more advanced techniques for explaining and visualizing the decision-making processes of AI models, such as counterfactual reasoning and concept activation maps, and can further enhance their transparency, trustworthiness, and clinical utility.

- *Federated learning*: This involves enabling the training of AI models on decentralized data across multiple institutions, while preserving privacy and security, and can accelerate the development and generalizability of CDSSs by utilizing larger, more diverse datasets.

- *Ambient intelligence*: This involves integrating AI-powered CDSSs with ubiquitous sensing and computing technologies, such as wearables, smart devices, and the Internet of Things, and can enable more continuous, context-aware, and proactive decision support in various care settings, from hospitals to homes.

It is crucial for researchers, practitioners, and policymakers to collaborate in addressing the technical, ethical, and organizational challenges, while using AI technologies to transform clinical decision-making and improve patient outcomes. By learning from the successes and failures of early adopters, and by engaging in multidisciplinary, patient-centered innovation, we can create more intelligent, personalized, and effective healthcare systems.

CHAPTER 9

The Road Ahead: What's Next for AI in Healthcare

The future of AI in healthcare is both exciting and challenging. As we have seen throughout this book, AI has the potential to revolutionize many aspects of healthcare delivery, from diagnosis and treatment to patient engagement and population health management. However, realizing this potential will require ongoing collaboration, innovation, and ethical reflection from a wide range of stakeholders, including clinicians, researchers, policymakers, and patients.

In this section, we explore some of the trends and opportunities that are shaping the future of AI in healthcare, as well as the challenges and considerations that will need to be addressed.

One of the most promising applications of AI in healthcare is in the area of precision medicine and personalized care. Precision medicine involves tailoring medical treatments and interventions to the specific characteristics and needs of individual patients, based on factors such as their genetic profile, medical history, lifestyle, and preferences (Ginsburg & Phillips, 2018).

AI can play a role in enabling precision medicine by analyzing large amounts of patient data to identify patterns and predict outcomes. For example, AI algorithms can be used to

- identify genetic variations associated with specific diseases or treatment responses
- predict the likelihood of disease progression or recurrence based on patient characteristics and biomarkers

- recommend personalized treatment plans based on a patient's unique clinical profile and preferences
- monitor patient responses to treatment in real-time and adjust interventions as needed

By enabling more precise and personalized care, AI has the potential to improve patient outcomes, reduce healthcare costs, and enhance the quality of life for patients with a wide range of conditions.

Population Health and Disease Prevention

Another opportunity for AI in healthcare is in the area of population health and disease prevention. Population health involves the proactive management of the health of entire communities or populations, with the goal of preventing disease and promoting wellness (Kindig & Stoddart, 2003).

AI can support population health efforts by analyzing large datasets to identify risk factors and predict disease outbreaks. For example, AI algorithms can be used to

- identify geographic areas or populations at high risk for specific diseases, such as diabetes or heart disease
- predict the spread of infectious diseases based on factors such as travel patterns, weather conditions, and social networks
- recommend targeted interventions or policy changes to prevent disease and promote health equity
- monitor the effectiveness of public health interventions and adjust strategies as needed

By enabling more proactive and targeted approaches to population health, AI has the potential to reduce the burden of chronic disease, improve health outcomes, and enhance the resilience of communities in the face of emerging health threats.

Challenges and Considerations

While the potential benefits of AI in healthcare are significant, there are also many challenges and considerations that need to be addressed as these technologies continue to evolve. Some of the challenges include the following:

- ensuring the safety and effectiveness of AI-based interventions, particularly in high-stakes clinical settings
- addressing bias and ensuring fairness in AI algorithms, particularly for underserved or marginalized populations
- protecting patient privacy and security in the context of large-scale data collection and analysis
- developing appropriate regulatory frameworks and standards for the development and deployment of AI in healthcare
- ensuring that AI-based interventions are accessible and affordable for all patients, regardless of their socioeconomic status or geographic location

To address these challenges, it is important for healthcare organizations, policymakers, and other stakeholders to engage in ongoing dialogue and collaboration. This may involve the following:

- developing interdisciplinary teams and partnerships that bring together expertise in AI, healthcare, ethics, and policy
- investing in research and development to advance the safety, effectiveness, and fairness of AI-based interventions
- engaging patients and communities in the design and implementation of AI-based solutions to ensure that they are responsive to their needs and preferences
- advocating for policies and regulations that promote the responsible and equitable development and deployment of AI in healthcare

Practical Tip: Healthcare organizations and clinicians should stay informed about the latest developments in AI and precision medicine, and consider how these technologies may be integrated into their practice. This may involve participating in research studies or pilot projects, attending conferences or workshops, and engaging in ongoing education and training.

As we consider the future of AI in healthcare, it is clear that there are both significant opportunities and challenges ahead. By working together to address these challenges and use these technologies, we have the potential to transform healthcare delivery and improve the lives of patients around the world. The future may be complex and uncertain, but with the right tools, partnerships, and mindset, we can succeed.

9.1 QUANTUM COMPUTING AND HEALTHCARE

One of the most exciting and potentially transformative technologies is quantum computing. Unlike classical computers, which rely on binary bits to process information, quantum computers use quantum bits (qubits) that can exist in multiple states simultaneously. This allows quantum computers to perform certain types of calculations much faster than classical computers, particularly in areas such as optimization, simulation, and machine learning (Biamonte et al., 2017).

In healthcare, quantum computing could have a wide range of applications, such as the following:

- accelerating drug discovery and development by simulating complex molecular interactions and predicting drug efficacy and safety (Cao et al., 2018)
- optimizing treatment plans and resource allocation by analyzing large datasets of patient outcomes and system performance (Chen et al., 2020)
- enhancing medical imaging and diagnostics by processing and analyzing high-dimensional data such as genomic sequences and brain scans (Li et al., 2018)
- improving cybersecurity and data privacy by enabling more secure encryption and communication protocols (Pirandola et al., 2020)

While quantum computing is still in its early stages, there are already several initiatives underway to explore its potential in healthcare.

Case Study: Cleveland Clinic's Quantum Computing Pilot

In 2019, Cleveland Clinic announced a partnership with IBM to explore the use of quantum computing in healthcare. As part of this partnership, Cleveland Clinic researchers are working with IBM's quantum computing experts to develop new algorithms and applications for drug discovery, patient care, and medical research.

One of the first projects under this partnership is focused on accelerating the discovery of new drugs for cancer and other diseases. Using IBM's quantum computing platform, Cleveland Clinic researchers are simulating the interactions between molecules and proteins to identify promising drug candidates more quickly and efficiently than traditional methods.

Another project is focused on optimizing patient care pathways and resource allocation. By analyzing large datasets of patient outcomes and system performance using quantum algorithms, Cleveland Clinic researchers hope to identify bottlenecks and inefficiencies in the healthcare system and develop more effective strategies for patient management.

While these projects are still in the early stages, they demonstrate the potential of quantum computing to transform healthcare delivery and research. As quantum computing technologies continue to advance, we can expect to see even more exciting applications emerge in the years to come.

Practical Tip: As quantum computing advances, healthcare organizations should start to explore its potential applications and implications for their specific contexts. This may involve partnering with academic institutions or technology companies to develop proof-of-concept projects, investing in workforce training and development, and engaging in scenario planning to anticipate and prepare for the impact of quantum computing on healthcare delivery and business models.

Quantum computing represents a new frontier in healthcare innovation, with the potential to accelerate discovery, optimize performance, and enhance security in ways that were previously unimaginable. As with any emerging technology, there will be challenges and uncertainties along the way, but the potential benefits are too significant to ignore. By staying informed, engaged, and proactive, healthcare organizations can utilize the power of quantum computing and shape the future of healthcare delivery for the better.

9.2 EDGE COMPUTING AND REAL-TIME ANALYSIS

Another trend in healthcare AI involves edge computing and real-time analysis. *Edge computing* refers to the practice of processing data locally, at or near the source of the data, rather than sending it to a centralized cloud or data center. This can enable faster, more efficient, and more secure data processing, particularly for applications that require real-time analysis and decision-making (Shi et al., 2016).

The healthcare industry generates massive amounts of data from various sources, including electronic health records, medical imaging, wearable devices, and sensor networks. Processing and analyzing this data in real time can provide valuable insights and support timely interventions, leading to improved patient outcomes and operational efficiency.

Edge computing has the potential to dramatically transform healthcare by processing and analyzing data closer to where it is generated, rather than solely relying on centralized cloud infrastructure. This distributed approach not only improves speed and efficiency but also helps address concerns around privacy, bandwidth, and real-time decision-making. Below are several ways that edge computing can be applied in healthcare, along with a discussion of the benefits it provides the following:

1. Real-time patient monitoring and analysis. By processing data directly on wearables or local gateways, edge computing makes it possible to detect and respond to clinical issues more rapidly. For example, wearable ECG patches, continuous glucose monitors, or pulse oximeters can analyze a patient's vitals in near real-time, sending alerts to clinicians the moment abnormal patterns arise (Etli et al., 2024). This immediacy helps in proactive interventions, potentially averting serious complications.

2. Telemedicine and remote care. Secure, low-latency communication is vital for telehealth services, especially when patients live in rural or underserved regions. Edge devices can encrypt and process data locally (e.g., video feed, vital signs), then transmit only the essential results to providers (Qiu et al., 2020). This approach conserves network bandwidth and creates a smoother experience for both clinicians and patients, ensuring that remote consultations and diagnostics proceed without significant delays or service interruptions.

3. Decentralized clinical trials and research. Traditionally, large-scale studies rely on complex data transfer and storage in centralized systems, which can slow down research and heighten security risks. With edge computing, sites hosting clinical trials can gather, preprocess, and analyze data locally before sending aggregated results to research centers (Satyanarayanan, 2017). This setup reduces the need for extensive infrastructure and enhances participant privacy by limiting the volume of raw data exchanged.

4. Heightened privacy and security. Healthcare data is inherently sensitive, and maintaining patient confidentiality is paramount. Edge computing enables local processing and secure storage of health information directly on devices or within a controlled network perimeter (Rahman et al., 2020). Since less data needs to traverse external networks or reside on distant servers, the risk of interception or unauthorized access decreases. Patients and providers thus have more confidence in how their data is managed, improving overall trust in digital health solutions.

Ultimately, edge computing architectures are well-suited to managing the growing volume, velocity, and variety of healthcare data. By reducing dependence on cloud-based processing, edge solutions help lower latency and bandwidth usage. This is particularly important for real-time or mission-critical applications, such as continuous patient monitoring or rapid emergency response. As healthcare ecosystems evolve to become more data-driven, edge computing stands out as a technology that can enhance responsiveness, bolster security, and ensure that innovations in patient care remain both practical and privacy-conscious.

Case Study: Edge AI for Early Alzheimer's Detection

Imagine developing an edge AI platform for early detection of Alzheimer's disease. The platform could utilize a combination of eye tracking and machine learning algorithms to assess cognitive function and identify early signs of cognitive decline.

Traditionally, cognitive assessments are performed in clinical settings using pen-and-paper tests or computerized assessments. However, these assessments can be time-consuming, expensive, and may not detect subtle changes in cognitive function that could indicate early stage Alzheimer's.

This platform uses a simple visual test that can be administered using a smartphone or tablet. During the test, the user's eye movements are tracked and analyzed using machine learning algorithms running on the device itself. This edge computing approach allows for real-time analysis of the user's cognitive function without the need for data to be sent to a centralized server.

Below is a sample code of how you might implement a project like this like this:

```python
import numpy as np
import tensorflow as tf
from dataclasses import dataclass
from typing import List, Dict, Tuple, Optional
from datetime import datetime
import cv2
from enum import Enum
```

```python
import json

class TestType(Enum):
    NOVEL_FAMILIAR = "novel_familiar"
    PATTERN_RECOGNITION = "pattern_recognition"
    SPATIAL_MEMORY = "spatial_memory"

@dataclass
class EyeTrackingData:
    timestamp: float
    x_position: float
    y_position: float
    pupil_size: float
    fixation_duration: float
    saccade_velocity: float

@dataclass
class TestResult:
    test_type: TestType
    start_time: datetime
    end_time: datetime
    eye_tracking_data: List[EyeTrackingData]
    recognition_scores: Dict[str, float]
    cognitive_metrics: Dict[str, float]

class EyeTracker:
    def __init__(self, camera_id: int = 0):
```

```python
        """Initialize eye tracking system using device camera."""
        self.camera = cv2.VideoCapture(camera_id)
        self.face_cascade = cv2.CascadeClassifier(
            cv2.data.haarcascades + 'haarcascade_frontalface_default.xml'
        )
        self.eye_cascade = cv2.CascadeClassifier(
            cv2.data.haarcascades + 'haarcascade_eye.xml'
        )

    def track_eyes(self, frame: np.ndarray) -> Optional[EyeTrackingData]:
        """Process a single frame to track eye movements."""
        gray = cv2.cvtColor(frame, cv2.COLOR_BGR2GRAY)
        faces = self.face_cascade.detectMultiScale(gray, 1.3, 5)

        for (x, y, w, h) in faces:
            roi_gray = gray[y:y+h, x:x+w]
            eyes = self.eye_cascade.detectMultiScale(roi_gray)

            if len(eyes) >= 2:
                # Calculate eye positions and movements
                eye_positions = self._calculate_eye_positions(eyes)
```

```python
                pupil_size = self._estimate_pupil_size(roi_gray, eyes)
                fixation, velocity = self._analyze_eye_movement(eye_positions)
                
                return EyeTrackingData(
                    timestamp=datetime.now().timestamp(),
                    x_position=eye_positions[0],
                    y_position=eye_positions[1],
                    pupil_size=pupil_size,
                    fixation_duration=fixation,
                    saccade_velocity=velocity
                )
        
        return None
    
    def _calculate_eye_positions(self, eyes: np.ndarray) -> Tuple[float, float]:
        """Calculate average eye position."""
        x_positions = [x + w/2 for (x, y, w, h) in eyes]
        y_positions = [y + h/2 for (x, y, w, h) in eyes]
        return np.mean(x_positions), np.mean(y_positions)
    
    def _estimate_pupil_size(self, roi_gray: np.ndarray, eyes: np.ndarray) -> float:
```

```python
        """Estimate pupil size from eye regions."""
        pupil_sizes = []
        for (ex, ey, ew, eh) in eyes:
            eye_roi = roi_gray[ey:ey+eh, ex:ex+ew]
            _, thresh = cv2.threshold(eye_roi, 30, 255, cv2.THRESH_BINARY_INV)
            contours, _ = cv2.findContours(thresh, cv2.RETR_TREE, cv2.CHAIN_APPROX_SIMPLE)
            if contours:
                pupil_sizes.append(cv2.contourArea(max(contours, key=cv2.contourArea)))
        return np.mean(pupil_sizes) if pupil_sizes else 0.0

    def _analyze_eye_movement(self, current_position: Tuple[float, float]) -> Tuple[float, float]:
        """Analyze eye movement patterns."""
        # In a real implementation, this would track movement over time
        # Here we return placeholder values
        return 0.2, 15.0  # fixation duration (s), saccade velocity (deg/s)


class CognitiveTest:
    def __init__(self, test_type: TestType):
        """Initialize cognitive test parameters."""
        self.test_type = test_type
        self.start_time = None
        self.end_time = None
```

```python
        self.eye_tracking_data = []

    def start_test(self):
        """Start a new cognitive test session."""
        self.start_time = datetime.now()
        self.eye_tracking_data = []

    def end_test(self) -> TestResult:
        """End the test and compute results."""
        self.end_time = datetime.now()

        # Calculate cognitive metrics based on eye tracking data
        recognition_scores = self._calculate_recognition_scores()
        cognitive_metrics = self._calculate_cognitive_metrics()

        return TestResult(
            test_type=self.test_type,
            start_time=self.start_time,
            end_time=self.end_time,
            eye_tracking_data=self.eye_tracking_data,
            recognition_scores=recognition_scores,
            cognitive_metrics=cognitive_metrics
        )
```

```python
    def add_eye_tracking_data(self, data: EyeTrackingData):
        """Add eye tracking data point to current test."""
        self.eye_tracking_data.append(data)

    def _calculate_recognition_scores(self) -> Dict[str, float]:
        """Calculate recognition scores based on eye tracking patterns."""
        if not self.eye_tracking_data:
            return {"novel": 0.0, "familiar": 0.0}

        # Analyze fixation patterns
        fixation_times = [data.fixation_duration for data in self.eye_tracking_data]
        saccade_velocities = [data.saccade_velocity for data in self.eye_tracking_data]

        # Calculate metrics
        avg_fixation = np.mean(fixation_times)
        avg_velocity = np.mean(saccade_velocities)

        # Novel vs familiar recognition scores
        novel_score = self._compute_novelty_score(avg_fixation, avg_velocity)
        familiar_score = 1.0 - novel_score

        return {
```

```python
                "novel": novel_score,
                "familiar": familiar_score
        }


    def _calculate_cognitive_metrics(self) -> Dict[str, float]:
        """Calculate cognitive metrics from eye tracking data."""
        if not self.eye_tracking_data:
            return {
                "attention_score": 0.0,
                "processing_speed": 0.0,
                "memory_score": 0.0
            }

        # Calculate attention score based on fixation patterns
        attention_score = np.mean([
            data.fixation_duration > 0.2 for data in self.eye_tracking_data
        ])

        # Calculate processing speed based on saccade velocities
        processing_speed = np.mean([
            data.saccade_velocity for data in self.eye_tracking_data
        ]) / 20.0  # Normalize to 0-1 scale
```

```python
        # Calculate memory score based on recognition patterns
        memory_score = self._compute_memory_score()

        return {
            "attention_score": attention_score,
            "processing_speed": processing_speed,
            "memory_score": memory_score
        }

    def _compute_novelty_score(self, avg_fixation: float, avg_velocity: float) -> float:
        """Compute novelty score based on eye movement patterns."""
        # Novel stimuli typically result in longer fixations and slower saccades
        fixation_factor = min(avg_fixation / 0.3, 1.0)  # Normalize to 0-1
        velocity_factor = max(1.0 - (avg_velocity / 30.0), 0.0)  # Normalize to 0-1
        return (fixation_factor + velocity_factor) / 2.0

    def _compute_memory_score(self) -> float:
        """Compute memory score based on recognition patterns."""
        if not self.eye_tracking_data:
            return 0.0
```

```python
        # Analyze pattern of fixations and saccades
        pattern_score = self._analyze_viewing_pattern()
        return pattern_score

    def _analyze_viewing_pattern(self) -> float:
        """Analyze viewing pattern for memory assessment."""
        if len(self.eye_tracking_data) < 2:
            return 0.0

        # Calculate pattern consistency
        positions = [(data.x_position, data.y_position) for data in self.eye_tracking_data]
        pattern_consistency = self._calculate_pattern_consistency(positions)
        return pattern_consistency

    def _calculate_pattern_consistency(self, positions: List[Tuple[float, float]]) -> float:
        """Calculate consistency of viewing pattern."""
        # Calculate average distance between consecutive positions
        distances = [
            np.sqrt((x2-x1)**2 + (y2-y1)**2)
            for (x1, y1), (x2, y2) in zip(positions[:-1], positions[1:])
        ]
```

```python
        # More consistent patterns have more uniform distances
        std_dev = np.std(distances)
        return max(1.0 - (std_dev / 100.0), 0.0)  # Normalize to 0-1


class EdgeAISystem:
    def __init__(self):
        """Initialize the Edge AI system."""
        self.eye_tracker = EyeTracker()
        self.ml_model = self._load_ml_model()

    def _load_ml_model(self) -> tf.keras.Model:
        """Load pre-trained ML model for cognitive assessment."""
        model = tf.keras.Sequential([
            tf.keras.layers.Dense(64, activation='relu', input_shape=(6,)),
            tf.keras.layers.Dropout(0.2),
            tf.keras.layers.Dense(32, activation='relu'),
            tf.keras.layers.Dense(16, activation='relu'),
            tf.keras.layers.Dense(1, activation='sigmoid')
        ])
        model.compile(optimizer='adam', loss='binary_crossentropy')
        return model
```

```python
    def run_cognitive_assessment(self, test_duration: int = 300) -> Dict:
        """Run a complete cognitive assessment."""
        test = CognitiveTest(TestType.NOVEL_FAMILIAR)
        test.start_test()

        start_time = datetime.now()
        while (datetime.now() - start_time).seconds < test_duration:
            ret, frame = self.eye_tracker.camera.read()
            if ret:
                eye_data = self.eye_tracker.track_eyes(frame)
                if eye_data:
                    test.add_eye_tracking_data(eye_data)

        # Get test results
        results = test.end_test()

        # Generate cognitive assessment
        assessment = self._generate_assessment(results)
        return assessment

    def _generate_assessment(self, results: TestResult) -> Dict:
```

```python
        """Generate cognitive assessment from test 
results."""
        # Prepare features for ML model
        features = np.array([
            np.mean([data.fixation_duration for data 
in results.eye_tracking_data]),
            np.mean([data.saccade_velocity for data in 
results.eye_tracking_data]),
            np.mean([data.pupil_size for data in 
results.eye_tracking_data]),
            results.
cognitive_metrics['attention_score'],
            results.
cognitive_metrics['processing_speed'],
            results.cognitive_metrics['memory_score']
        ]).reshape(1, -1)

        # Get cognitive decline risk score
        risk_score = float(self.ml_model.
predict(features)[0])

        return {
            "test_type": results.test_type.value,
            "duration": (results.end_time - results.
start_time).seconds,
            "cognitive_metrics": results.
cognitive_metrics,
            "recognition_scores": results.
recognition_scores,
```

```python
                "risk_score": risk_score,
                "assessment": self._interpret_risk_score(risk_score),
                "recommendations": self._generate_recommendations(risk_score)
            }

    def _interpret_risk_score(self, risk_score: float) -> str:
        """Interpret the risk score."""
        if risk_score < 0.2:
            return "Low risk - No significant cognitive decline detected"
        elif risk_score < 0.4:
            return "Mild risk - Some subtle changes in cognitive patterns"
        elif risk_score < 0.6:
            return "Moderate risk - Notable changes in cognitive function"
        else:
            return "High risk - Significant cognitive changes detected"

    def _generate_recommendations(self, risk_score: float) -> List[str]:
        """Generate recommendations based on risk score."""
        recommendations = ["Continue regular cognitive assessments"]
```

```python
        if risk_score >= 0.4:
            recommendations.extend([
                "Schedule comprehensive neurological evaluation",
                "Consider cognitive enhancement activities",
                "Monitor for changes in daily activities"
            ])

        if risk_score >= 0.6:
            recommendations.extend([
                "Urgent consultation with neurologist",
                "Begin early intervention protocols",
                "Implement daily cognitive monitoring"
            ])

        return recommendations

# Example usage
if __name__ == "__main__":
    system = EdgeAISystem()

    print("Starting cognitive assessment...")
    results = system.run_cognitive_assessment(test_duration=30)  # Shorter duration for demo
```

```
    print("\nAssessment Results:")
    print(f"Test Type: {results['test_type']}")
    print(f"Duration: {results['duration']} seconds")
    print("\nCognitive Metrics:")
    for metric, value in results['cognitive_metrics'].
items():
        print(f"- {metric}: {value:.2f}")
    print(f"\nRisk Score: 
{results['risk_score']:.2f}")
    print(f"Assessment: {results['assessment']}")
    print("\nRecommendations:")
    for rec in results['recommendations']:
        print(f"- {rec}")
```

Enabling early detection and intervention, this edge AI platform has the potential to improve outcomes for patients with Alzheimer's and reduce the burden on healthcare systems.

Recent Research: A 2021 study by Nguyen et al. proposed a federated learning framework for edge computing in healthcare IoT systems. The framework enables distributed training of machine learning models across multiple edge devices while preserving data privacy and reducing communication overhead. The authors demonstrated the effectiveness of the approach using a case study of fall detection in elderly patients (Nguyen et al., 2021).

As the healthcare industry continues to generate vast amounts of data, edge computing and real-time analysis will play an increasingly important role in enabling timely insights and interventions. By bringing computation closer to the data source, edge computing can help address the challenges of latency, bandwidth, privacy, and security in healthcare AI applications. As edge computing technologies continue to evolve, we can expect to see more innovative applications emerge that transform the way we deliver and experience healthcare.

9.2.1 5G Networks and Telemedicine

The rollout of 5G networks is another trend that is expected to have a significant impact on healthcare AI. 5G networks offer much faster speeds, lower latency, and higher capacity than previous generations of wireless networks, enabling a wide range of new applications and services (Agiwal et al., 2016).

In healthcare, 5G networks could be particularly transformative for telemedicine and remote patient monitoring. Telemedicine involves the use of telecommunications technologies to provide remote medical care, consultation, and education. Remote patient monitoring involves the use of sensors and devices to collect and transmit patient data to healthcare providers for analysis and intervention.

With 5G, patients and providers could have access to high-quality video consultations, remote diagnostic imaging, and real-time monitoring of vital signs and other health data, regardless of location. This could help to improve access to care for underserved and rural populations, reduce healthcare costs, and enable more personalized and proactive care delivery (Ahad et al., 2019).

Some of the benefits of 5G for telemedicine and remote patient monitoring include the following:

- higher bandwidth and lower latency, enabling high-quality video consultations and real-time data transmission
- increased reliability and availability, ensuring that critical health data is transmitted quickly and accurately
- enhanced security and privacy, protecting sensitive patient data from unauthorized access or tampering
- greater scalability and flexibility, allowing for the deployment of telemedicine and remote monitoring solutions in a variety of settings and contexts

9.3 FDA GUIDANCE ON AI/ML-BASED SOFTWARE

As AI becomes more prevalent in healthcare, regulatory agencies are beginning to develop guidance and frameworks for ensuring the safety, efficacy, and fairness of AI-based medical devices and software. In the United States, the Food and Drug Administration (FDA) has issued several guidance documents on the regulation of AI/ML-based software as a medical device (SaMD) (FDA, 2019).

The FDA's guidance emphasizes the importance of transparency, accountability, and continuous monitoring in the development and deployment of AI/ML-based SaMD. Important principles include the following:

- establishing clear intended uses and performance metrics for AI/ML-based SaMD, and validating the software's performance against these metrics
- ensuring that the data used to train and validate AI/ML-based SaMD is representative of the intended patient population and use case
- implementing robust cybersecurity and data privacy controls to protect patient data and prevent unauthorized access or misuse
- providing clear labeling and user instructions for AI/ML-based SaMD, including information on the software's intended use, performance, and limitations
- establishing processes for continuous monitoring and updating of AI/ML-based SaMD, including mechanisms for detecting and mitigating potential biases or errors

The FDA's guidance also recognizes the unique challenges posed by AI/ML-based SaMD, such as the potential for the software to adapt and evolve over time based on new data inputs. To address these challenges, the FDA has proposed a regulatory framework that includes a "predetermined change control plan" that specifies the types of modifications that can be made to the software without requiring additional regulatory review.

While the FDA's guidance is an important step toward creating a regulatory framework for AI in healthcare, there is still much work to be done to ensure that AI-based medical devices and software are safe, effective, and equitable. Some of the key challenges and considerations include the following:

- developing standardized metrics and benchmarks for evaluating the performance and safety of AI/ML-based SaMD
- ensuring that AI/ML-based SaMD is developed and validated using diverse and representative datasets to minimize bias and ensure fairness
- balancing the need for continuous monitoring and updating of AI/ML-based SaMD with the need for regulatory oversight and patient safety
- engaging with patients, providers, and other stakeholders to ensure that AI/ML-based SaMD is developed and deployed in a way that is transparent, accountable, and aligned with patient needs and values

Practical Tip: Healthcare organizations developing or deploying AI/ML-based software should familiarize themselves with the FDA's guidance and ensure that their practices align with the key principles outlined therein. This may involve establishing clear governance structures, implementing robust testing and validation processes, and engaging with regulators and other stakeholders to ensure transparency and accountability.

As AI continues to evolve and become more integrated into healthcare delivery, it will be essential for regulatory frameworks to keep pace and ensure that these technologies are developed and deployed in a way that prioritizes patient safety, efficacy, and equity. The FDA's guidance on AI/ML-based SaMD is an important step in this direction, but there is still much work to be done to build a comprehensive and effective regulatory framework for AI in healthcare.

9.3.1 GDPR and International Data Protection Laws

Another regulatory issue in healthcare AI is data protection and privacy. With the increasing collection and use of patient data for AI applications, there is a growing concern about the potential for misuse or unauthorized access to sensitive health information.

In the European Union, the General Data Protection Regulation (GDPR) sets strict requirements for the collection, use, and storage of personal data, including health data (European Parliament and Council of European Union, 2016). Under GDPR, healthcare organizations must obtain explicit consent from patients for the use of their data in AI applications, and must provide clear information about how the data will be used and protected. GDPR also gives patients the right to access, correct, and delete their data, and requires organizations to implement appropriate technical and organizational measures to ensure data security and privacy.

Other countries and regions are also developing their own data protection laws and regulations, which may have implications for healthcare AI. For example, in the United States, the Health Insurance Portability and Accountability Act (HIPAA) sets standards for the protection of patient health information, while the California Consumer Privacy Act (CCPA) gives consumers the right to know what personal information is being collected about them and how it is being used.

Navigating this complex landscape of data protection laws and regulations can be challenging for healthcare organizations developing or deploying AI applications. Some of the considerations include the following:

- ensuring that patient data is collected, used, and stored in compliance with applicable laws and regulations
- obtaining explicit consent from patients for the use of their data in AI applications, and providing clear information about how the data will be used and protected
- implementing appropriate technical and organizational measures to ensure data security and privacy, such as encryption, access controls, and monitoring systems
- establishing clear policies and procedures for responding to data breaches or other security incidents
- providing patients with access to their data and the ability to correct or delete it as required by law

As healthcare AI continues to evolve, it will be important for organizations to stay up-to-date with the latest data protection laws and regulations, and to implement appropriate measures to ensure the privacy and security of patient data.

Recent Research: A 2021 study by Favaretto et al. explored the implications of GDPR for the development and deployment of AI in healthcare. The authors discussed the challenges of obtaining valid consent for AI applications, ensuring data minimization and purpose limitation, and providing meaningful explanations of AI decision-making. They also proposed a framework for "privacy-preserving AI" that utilizes techniques such as federated learning and differential privacy to enable the development of AI models while protecting patient privacy (Favaretto et al., 2019).

As the use of AI in healthcare continues to grow, it will be essential for organizations to prioritize data protection and privacy in the development and deployment of these technologies. By implementing appropriate technical and organizational measures, obtaining explicit consent from patients, and staying up-to-date with the latest data protection laws and regulations, healthcare organizations can unlock the potential of AI while ensuring the privacy and security of patient data.

9.4 REIMBURSEMENT MODELS AND INCENTIVE STRUCTURES

Another policy issue in healthcare AI is reimbursement and incentive structures. Currently, many healthcare systems are based on fee-for-service models, which reward providers for the quantity of services delivered rather than the quality of outcomes achieved. This can create disincentives for the adoption of AI and other technologies that may improve outcomes but reduce the volume of services delivered.

To address this issue, some healthcare organizations and policymakers are exploring new reimbursement models and incentive structures that align with the goals of value-based care. *Value-based care* is a healthcare delivery model that focuses on improving patient outcomes and reducing costs by incentivizing providers to deliver high-quality, cost-effective care (Porter & Teisberg, 2006).

Some examples of value-based reimbursement models and incentive structures that could support the adoption of AI in healthcare include the following:

- Bundled payments: Under this model, providers receive a fixed payment for a specific episode of care, such as a surgery or a chronic disease management program. This can incentivize providers to use AI and other technologies to optimize care delivery and reduce complications and readmissions.

- Accountable care organizations (ACOs): ACOs are groups of healthcare providers that work together to coordinate care for a defined patient population. ACOs are incentivized to improve quality and reduce costs through shared savings and risk arrangements, which can encourage the adoption of AI and other technologies that support population health management.

- Pay-for-performance (P4P) programs: P4P programs provide financial incentives to providers for meeting specific quality and cost metrics, such as reducing hospital-acquired infections or improving patient satisfaction. AI and other technologies that support quality improvement and cost reduction could be incentivized under P4P programs.

- Value-based purchasing (VBP): VBP is a strategy used by payers, such as Medicare and Medicaid, to link provider payments to quality and cost outcomes. VBP programs can create incentives for providers to adopt AI and other technologies that improve outcomes and reduce costs.

Implementing these value-based reimbursement models and incentive structures can be challenging, as they require significant changes to existing payment systems and care delivery models. Some of the considerations for policymakers and healthcare organizations include the following:

- defining and measuring quality and cost outcomes in a way that is meaningful and fair to providers and patients
- ensuring that value-based payment models do not create unintended consequences, such as underutilization of necessary services or cherry-picking of healthier patients
- providing technical assistance and support to providers to help them adopt AI and other technologies that support value-based care
- engaging patients and other stakeholders in the design and implementation of value-based payment models to ensure that they are patient-centered and equitable

Practical Tip: Healthcare organizations and policymakers should work together to develop and test new reimbursement models and incentive structures that support the adoption of AI and other technologies that improve outcomes and reduce costs. This may involve piloting new payment models in specific geographic regions or patient populations, engaging with payers and other stakeholders to negotiate value-based contracts, and providing education and support to providers to help them navigate the transition to value-based care.

As the healthcare industry continues to shift toward value-based care, it will be essential to align reimbursement and incentive structures with the goals of improving patient outcomes and reducing costs. By creating financial incentives for the adoption of AI and other technologies that support these goals, healthcare organizations and policymakers can accelerate the transformation of the healthcare system and improve the lives of patients.

9.5 FEDERATED LEARNING AND PRIVACY-PRESERVING AI

One of the challenges in healthcare AI is the need to train models on large, diverse datasets while protecting patient privacy and data security. *Federated learning* is an emerging approach that addresses this challenge by enabling AI models to be trained on decentralized datasets without the need for data sharing or centralization (Rieke et al., 2020).

In federated learning, each participating institution or device trains a local AI model on its own dataset, and then shares only the model parameters (not the raw data) with a central server. The central server aggregates the parameters from multiple local models to create a global model, which is then distributed back to the local devices for further training. This process is repeated iteratively until the global model converges to a satisfactory level of performance (Yang et al., 2019).

Federated learning has several advantages for healthcare AI, including the following:

- enabling the creation of large, diverse datasets for AI training while preserving patient privacy and data security
- reducing the need for data sharing and centralization, which can be costly and time-consuming
- allowing for the creation of personalized AI models that are tailored to the specific needs and characteristics of individual patients or populations

However, implementing federated learning in healthcare also presents several challenges, such as the following:

- ensuring the quality and representativeness of the local datasets used for training
- dealing with the heterogeneity and variability of the local datasets and models
- communicating and aggregating the model parameters efficiently and securely
- validating and governing the global model to ensure its safety, fairness, and effectiveness

To address these challenges, researchers and practitioners are exploring various techniques and frameworks for privacy-preserving AI, such as the following:

- differential privacy: a mathematical framework for adding noise to the model parameters to protect individual privacy while still allowing for useful insights to be learned from the data (Dwork & Roth, 2014)
- secure multi-party computation: a cryptographic protocol that allows multiple parties to jointly compute a function over their inputs while keeping those inputs private (Cramer et al., 2015)

- homomorphic encryption: A type of encryption that allows computations to be performed on encrypted data without decrypting it first, enabling secure data aggregation and analysis (Gentry, 2009)

Case Study: MELLODDY (Machine Learning Ledger Orchestration for Drug Discovery)

MELLODDY is a consortium of ten pharmaceutical companies and seven technology partners that aims to use federated learning to accelerate drug discovery while protecting the confidentiality of each company's proprietary data. The consortium has developed a platform that allows each company to train its own AI models on its own datasets, and then share only the model parameters with a central server for aggregation and optimization.

The MELLODDY platform uses a combination of federated learning, differential privacy, and secure multi-party computation to ensure the privacy and security of the participating companies' data. The platform also includes a blockchain-based ledger to track the provenance and usage of the models and datasets, and to ensure the fair attribution of intellectual property rights.

In a pilot study, the MELLODDY platform was used to train a global model for predicting the toxicity of chemical compounds, using datasets from multiple pharmaceutical companies. The study found that the federated model achieved similar performance to a centralized model trained on the same data, while preserving the confidentiality of each company's proprietary data (Heyndrickx et al., 2024).

The MELLODDY platform demonstrates the potential of federated learning and privacy-preserving AI to enable collaborative and decentralized AI development in healthcare, while protecting patient privacy and data security.

Practical Tip: When implementing federated learning in healthcare AI, it is important to carefully consider the design of the federated learning architecture, including the choice of aggregation algorithms, the frequency of model updates, and the mechanisms for ensuring data privacy and security. Engaging with experts in cryptography, distributed systems, and privacy law can help to ensure that the federated learning system is robust, secure, and compliant with relevant regulations.

Federated learning and privacy-preserving AI represent promising approaches for enabling the development of large-scale, collaborative AI models in healthcare, while protecting patient privacy and data security. As these technologies continue to mature, they have the potential to accelerate the pace of innovation in healthcare AI, while ensuring that the benefits of these technologies are realized in an ethical and responsible manner.

9.6 EXPLAINABLE AI AND INTERPRETABLE MODELS

Another research frontier in healthcare AI is the development of explainable and interpretable models. Many current AI models, particularly those based on deep learning, are often seen as black boxes that make predictions or decisions without providing clear explanations or rationales. This lack of transparency can be problematic in healthcare, where understanding the reasoning behind AI-based recommendations is crucial for building trust and ensuring patient safety (Amann et al., 2020).

Explainable AI (XAI) is an emerging field that aims to create AI models that are more transparent, interpretable, and accountable. XAI techniques can help to

- provide clear explanations for how AI models make predictions or decisions, in terms that are understandable to healthcare providers and patients
- identify potential biases or errors in AI models, and provide mechanisms for correcting or mitigating them
- enable healthcare providers to integrate AI recommendations with their own clinical judgment and expertise
- facilitate regulatory oversight and auditing of AI models to ensure their safety and effectiveness

There are several approaches to developing explainable and interpretable AI models, including the following:

- Rule-based systems: These models use a set of explicit, human-readable rules to make decisions or predictions. While they may be less accurate than more complex models, they are often more transparent and interpretable.
- Decision trees and random forests: These models use a series of binary decisions to arrive at a prediction or recommendation. They can be visualized as a tree-like structure, making them relatively easy to interpret.

- Linear and logistic regression: These statistical models find a linear relationship between input features and output variables. The coefficients of the model can be interpreted as the relative importance of each feature.
- Attention mechanisms: These techniques allow neural networks to focus on specific parts of the input data when making predictions. By visualizing the attention weights, we can gain insights into what the model is "paying attention to" when making a decision.
- Post hoc explanations: These methods aim to provide explanations for the decisions of a pretrained black box model. Techniques such as LIME (Local Interpretable Model-Agnostic Explanations) and SHAP (SHapley Additive exPlanations) can help to identify the most important features for a particular prediction.

While these approaches can help to improve the interpretability of AI models, there is often a trade-off between interpretability and accuracy. More complex models, such as deep neural networks, may achieve higher accuracy but be more difficult to interpret. Finding the right balance between performance and explainability is an ongoing challenge in healthcare AI research.

Case Study: Explainable AI for Breast Cancer Diagnosis

Researchers at the Massachusetts Institute of Technology (MIT) have developed an explainable AI system for breast cancer diagnosis that aims to provide clear visualizations of the features that contribute to a particular diagnosis (Shen et al., 2019). The system uses a combination of deep learning and attention mechanisms to analyze mammogram images and identify regions of interest that are most relevant to the diagnosis.

In addition to providing a binary diagnosis (benign or malignant), the system also generates a saliency map that highlights the regions of the image that were most important for the diagnosis. This allows radiologists to see what the AI system is "paying attention to" and to compare its reasoning with their own clinical judgment.

In a study of 335 mammogram images, the explainable AI system achieved a diagnostic accuracy of 94%, which was comparable to that of experienced radiologists. The saliency maps generated by the system were also found to align well with the regions of interest identified by the radiologists.

This case study demonstrates the potential of explainable AI to provide transparent and interpretable decision support for medical diagnosis, while still achieving high levels of accuracy. By providing clear visualizations of the reasoning behind AI-based recommendations, explainable AI systems can help to build trust and facilitate collaboration between human and machine intelligence in healthcare.

Recent Research: A 2021 study by Payrovnaziri et al. proposed an explainable AI framework for predicting mortality risk in intensive care unit patients. The framework combined deep learning with gradient boosting and provided visualizations of the key features contributing to each prediction, enabling clinicians to understand and validate the model's reasoning. The authors found that the explainable model achieved comparable performance to a black box deep learning model while providing greater transparency and interpretability.

As AI becomes increasingly integrated into healthcare decision-making, the development of explainable and interpretable models will be essential for ensuring the safety, effectiveness, and trustworthiness of these systems. By providing clear insights into the reasoning behind AI-based recommendations, explainable AI can help to foster collaboration between human and machine intelligence, and ultimately improve patient outcomes.

9.7 MULTIMODAL LEARNING AND SENSOR FUSION

Another promising research direction in healthcare AI is multimodal learning and sensor fusion. Multimodal learning involves the integration of multiple types of data, such as images, text, and audio, into a single AI model. Sensor fusion involves the combination of data from multiple sensors or devices to create a more comprehensive and accurate picture of a patient's health status (Ramachandram & Taylor, 2017).

Multimodal learning and sensor fusion have several potential applications in healthcare AI, including the following:

- enabling more accurate and comprehensive diagnosis and monitoring of complex health conditions, such as cancer or cardiovascular disease
- facilitating the integration of data from multiple sources, such as electronic health records, wearable devices, and patient-reported outcomes
- supporting the development of personalized treatment plans that take into account a patient's unique characteristics and preferences
- enhancing the accuracy and efficiency of clinical decision support systems by providing a more complete and nuanced view of patient data

One of the challenges in multimodal learning and sensor fusion is dealing with the heterogeneity and variability of the different data types and sources. Each modality may have its own unique characteristics, such as sampling rate, resolution, and noise level, which can make it difficult to combine them effectively.

To address this challenge, researchers are exploring various techniques for multimodal data integration and fusion, such as the following:

- deep learning architectures that can learn joint representations of multiple modalities, such as convolutional neural networks (CNNs) for images and recurrent neural networks (RNNs) for time series data
- attention mechanisms that can selectively focus on the most relevant features or modalities for a particular task or patient
- transfer learning approaches that can leverage pretrained models or features from one modality to improve the performance of another
- probabilistic graphical models that can capture the dependencies and uncertainties between different modalities and sensors

Case Study: Multimodal AI for Alzheimer's Disease Diagnosis

Researchers at the University of California, San Francisco (UCSF) have developed a multimodal AI system for the early diagnosis of Alzheimer's disease that combines data from multiple sources, including magnetic resonance imaging (MRI), positron emission tomography (PET), and cerebrospinal fluid (CSF) biomarkers (Zhang et al., 2019).

The system uses a deep learning architecture called a multimodal convolutional neural network (MCNN) to learn joint representations of the different data types. The MCNN consists of separate convolutional layers for each modality, followed by fully connected layers that combine the features from each modality.

In a study of 1,921 patients from the Alzheimer's Disease Neuroimaging Initiative (ADNI) database, the multimodal AI system achieved an accuracy of 91% in distinguishing between patients with Alzheimer's disease, mild cognitive impairment, and healthy controls. This was significantly higher than the accuracy of any individual modality alone, demonstrating the value of multimodal data integration.

The UCSF researchers also used an attention mechanism to identify the most important features and regions for the diagnosis, providing interpretability and transparency to the AI system's decision-making process. This allowed clinicians to compare the AI system's reasoning with their own clinical judgment and to identify potential areas for further investigation or intervention.

This case study reveals the potential of multimodal learning and sensor fusion to enable more accurate and comprehensive diagnosis of complex health conditions, while also providing interpretability and transparency to the AI decision-making process.

Recent Research: A 2021 study by Pang et al. proposed a multimodal AI framework for predicting the risk of cardiovascular disease that combines data from electronic health records, wearable devices, and social media. The framework uses a hierarchical attention network to selectively focus on the most relevant features and modalities for each patient, and achieves an area under the receiver operating characteristic curve (AUC) of 0.89, outperforming traditional risk prediction models (Pang et al., 2021).

As the volume and diversity of healthcare data continue to grow, the development of multimodal learning and sensor fusion techniques will be essential for unlocking the full potential of this data to improve patient outcomes. By enabling the integration and analysis of multiple data types and sources, these techniques can provide a more comprehensive and nuanced view of patient health, and support the development of personalized and precision medicine approaches.

CHAPTER 10

RESOURCES AND CONTINUING EDUCATION

10.1 THE IMPORTANCE OF CONTINUOUS LEARNING

Artificial intelligence (AI) is in a state of constant change in healthcare, and continuous learning is not just beneficial, it is critical. As new technologies, methodologies, and applications emerge at an unprecedented pace, clinicians, researchers, and informatics professionals must keep learning.

Continuous learning enables you to do the following:

1. Stay relevant: By keeping abreast of the latest developments, you can ensure that your skills and knowledge remain relevant and valuable in the ever-changing landscape of healthcare AI.

2. Improve patient care: As AI technologies advance, they offer new opportunities to enhance diagnostic accuracy, personalize treatments, and optimize outcomes. Continuous learning allows you to utilize these innovations to provide better care for your patients.

3. Advance your career: Demonstrating a commitment to ongoing education and staying current with the latest discoveries in the field can lead to new career opportunities, leadership roles, and research collaborations.

4. Contribute to the field: By engaging in continuous learning, you can identify gaps in current knowledge, generate new ideas, and contribute to the advancement of AI in healthcare through research, innovation, and knowledge sharing.

5. Adapt to changing regulations: As the use of AI in healthcare expands, so too does the regulatory landscape. Continuous learning helps you stay informed about evolving guidelines, standards, and best practices to ensure compliance and ethical use of AI technologies.

In the following sections, we explore strategies for effective continuous learning and guide you in curating a personalized learning plan to help you stay at the forefront of this exciting field.

10.2 STRATEGIES FOR KEEPING UP-TO-DATE

Keeping pace with advancements in healthcare AI can seem daunting, but by employing a combination of strategies, you can effectively stay up-to-date and integrate continuous learning into your professional life. Consider the following approaches:

1. Set aside dedicated learning time: Block out regular time slots in your schedule for reading, researching, and exploring new topics related to AI in healthcare. Treat this time as a priority, just as you would any other important commitment.

2. Use online resources: Take advantage of the online resources available, such as webinars, podcasts, blogs, and e-learning platforms. Many of these resources are free or low-cost and offer flexible learning opportunities that can fit into your busy schedule.

3. Engage with professional communities: Join online forums, social media groups, and professional organizations dedicated to healthcare AI. These communities provide valuable opportunities to network with peers, share knowledge, and learn from experts in the field.

4. Attend conferences and workshops: Participate in conferences, workshops, and seminars focused on AI in healthcare. These events offer concentrated learning experiences, exposure to cutting-edge research, and opportunities to connect with thought leaders and innovators.

5. Pursue formal education and certification: Consider enrolling in university courses, graduate programs, or certification courses to gain a deeper understanding of AI and its applications in healthcare. Many institutions now offer specialized programs tailored to the needs of healthcare professionals.

6. Collaborate with interdisciplinary teams: Engage in collaborative projects with professionals from diverse backgrounds, such as data scientists, engineers, and informaticians. Working alongside experts in other fields can broaden your perspective and expose you to new ideas and approaches.

7. Embrace hands-on learning: Seek out opportunities to apply your knowledge through practical projects, such as developing AI algorithms, participating in research studies, or implementing AI tools in your clinical practice. Hands-on experience is invaluable for deepening your understanding and staying current with emerging technologies.

By incorporating a mix of these strategies into your professional development plan, you can create a sustainable and effective approach to continuous learning in the field of healthcare AI.

10.3 CURATING YOUR INFORMATION SOURCES

With the abundance of information available on AI in healthcare, it is essential to curate a reliable and relevant set of sources to support your continuous learning journey. Here are some tips for selecting and managing your information sources:

1. Identify thought leaders and influencers: Follow the work of recognized experts, researchers, and innovators in the field. Their publications, presentations, and social media posts can provide valuable insights and keep you informed about the latest developments.

2. Subscribe to reputable journals and publications: Regularly read peer-reviewed journals and publications that focus on AI in healthcare, such as the *Journal of the American Medical Informatics Association* (JAMIA), *IEEE Journal of Biomedical and Health Informatics*, and *npj Digital Medicine*.

3. Utilize academic search engines and databases: Use tools like Google Scholar, PubMed, and IEEE Xplore to find and access scholarly articles, conference papers, and research studies related to your areas of interest.

4. Curate a list of trusted websites and blogs: Identify reputable websites and blogs that cover healthcare AI news, trends, and best practices. Regularly visit these sites or subscribe to their newsletters to stay informed about the latest developments.

5. Utilize social media and professional networks: Follow relevant hashtags, join LinkedIn groups, and participate in Twitter chats related to AI in healthcare. These platforms can help you discover new resources, engage in discussions, and stay connected with the broader community.

6. Use content aggregation and curation tools: Employ tools like RSS feeds, Google Alerts, and content curation platforms (e.g., Feedly and Flipboard) to centralize and organize information from multiple sources, making it easier to stay on top of new content.

7. Regularly review and update your sources: Periodically assess the relevance and quality of your information sources. Remove outdated or less relevant sources, and actively seek out new ones to ensure you have a diverse and up-to-date collection of resources.

Curating your information sources and developing a systematic approach to consuming and managing content will aid in creating a solid foundation for your continuous learning efforts in the rapidly evolving field of AI in healthcare.

10.4 RECOMMENDED READING AND REFERENCE MATERIALS

Here is a list of recommended reading materials and reference resources. These books, journals, and online publications offer valuable insights into the fundamentals of AI in healthcare, current research and applications, and future directions for the field.

10.4.1 Important Journals and Publications

1. *Journal of the American Medical Informatics Association* (JAMIA). JAMIA is a peer-reviewed journal that publishes research articles, reviews, and commentaries on the application of informatics in health and biomedicine. It frequently features studies on AI and machine learning in healthcare.

2. *IEEE Journal of Biomedical and Health Informatics.* This journal focuses on the application of information technology and computational methods in healthcare and biomedicine. It publishes research articles on topics such as AI, data mining, and decision support systems.

3. *npj Digital Medicine.* This open-access journal publishes research articles and reviews on the use of digital technologies, including AI, in healthcare. It covers a wide range of topics, from digital biomarkers and remote monitoring to personalized medicine and ethics.

10.4.2 Online Articles and Blogs

1. *Healthcare IT News.* This online publication covers the latest news, trends, and innovations in healthcare technology, including AI and machine learning. It features articles, interviews, and case studies from healthcare organizations and technology vendors.

2. *The Medical Futurist.* Founded by Dr. Bertalan Mesko, *The Medical Futurist* is a blog and website that explores the future of medicine and healthcare. It features articles, podcasts, and videos on topics such as AI, digital health, and personalized medicine.

3. *AI in Healthcare on Medium.* Medium is an online publishing platform that hosts a variety of publications and blogs, including *AI in Healthcare*. This publication features articles and stories from healthcare professionals, researchers, and technology experts on the application of AI in healthcare.

These recommended reading materials and reference resources can deepen your understanding of AI in healthcare, help you stay informed about the latest research and trends, and gain practical insights to support your continuous learning efforts.

10.5 CONFERENCES AND PROFESSIONAL ORGANIZATIONS

Attending conferences and engaging with professional organizations are valuable ways to stay current with the latest developments in AI and healthcare, network with peers and experts, and contribute to the advancement of the field. Here are some notable conferences and organizations to consider:

10.5.1 Major Healthcare AI Conferences

1. HIMSS Global Health Conference and Exhibition: The Healthcare Information and Management Systems Society (HIMSS) hosts this annual conference, which brings together healthcare professionals, technology vendors, and thought leaders to discuss the latest innovations in healthcare IT, including AI and machine learning.

2. AI in Healthcare Summit: This annual summit focuses specifically on the application of AI in healthcare, covering topics such as clinical decision support, drug discovery, and personalized medicine. It features keynote speeches, panel discussions, and networking opportunities with industry leaders and researchers.

3. Intelligent Health AI: This global summit series brings together healthcare professionals, AI experts, and policymakers to discuss the latest advancements and challenges in applying AI to healthcare. It features presentations, workshops, and exhibitions showcasing cutting-edge AI technologies and solutions.

10.5.2 Clinical Informatics Societies

1. American Medical Informatics Association (AMIA). AMIA is a professional organization dedicated to the science and practice of informatics in healthcare. It offers educational programs, publications, and networking opportunities for informatics professionals, including those focused on AI and machine learning.

2. Healthcare Information and Management Systems Society (HIMSS). HIMSS is a global organization focused on the use of information and technology in healthcare. It provides educational resources, professional development opportunities, and advocacy for the adoption of innovative technologies, including AI, in healthcare.

3. International Medical Informatics Association (IMIA). IMIA is a global organization that promotes the development and application of informatics in healthcare. It brings together national informatics associations, academic institutions, and industry partners to advance the field of medical informatics, including research and education on AI in healthcare.

10.5.3 Online Forums and Communities

1. Healthcare AI Subreddit. Reddit is a popular online platform that hosts a variety of communities, known as subreddits, dedicated to specific topics. The healthcare AI subreddit is a community of healthcare professionals, researchers, and enthusiasts who share news, discuss ideas, and ask questions related to AI in healthcare.

2. AI in Healthcare LinkedIn Group. LinkedIn is a professional networking platform that also hosts various groups focused on specific industries

or topics. The AI in Healthcare group is a community of professionals interested in the application of AI in healthcare, offering opportunities to share insights, ask questions, and connect with peers.

3. HealthcareAI Slack Community. Slack is a messaging and collaboration platform that enables the creation of communities, known as workspaces, for specific topics or interests. The HealthcareAI Slack community is a workspace dedicated to discussing the use of AI in healthcare, featuring channels for various subtopics and opportunities for real-time engagement with other members.

By participating in these conferences, professional organizations, and online communities, you can stay connected with the broader healthcare AI community, learn from experts and peers, and contribute to the ongoing dialogue and advancement of the field.

10.6 TRAINING AND CERTIFICATION PROGRAMS

To develop a deeper understanding of AI in healthcare and acquire specialized skills, consider pursuing training and certification programs. These programs can help you gain the knowledge and credentials needed to advance your career and make meaningful contributions to the field.

10.6.1 University Courses and Degrees

1. Master of Science in Clinical Informatics. Many universities now offer master's degree programs in clinical informatics, which cover topics such as data analytics, decision support systems, and AI in healthcare. These programs are designed for healthcare professionals looking to specialize in informatics and typically require a bachelor's degree and relevant work experience.

2. Graduate Certificate in Health Informatics. For those who may not have the time or resources to pursue a full master's degree, graduate certificate programs in health informatics offer a shorter, more focused alternative. These programs cover the essential concepts and skills in informatics, including AI and machine learning, and can often be completed in one year or less.

3. PhD in Biomedical Informatics. For those interested in pursuing a career in research or academia, a PhD in biomedical informatics provides

advanced training in the theories, methods, and applications of informatics in healthcare. These programs typically involve coursework, research projects, and a dissertation focused on a specific area of informatics, such as AI or data science.

10.6.2 Online Learning Platforms

1. Coursera: AI in Healthcare Specialization Coursera is an online learning platform that partners with top universities and organizations to offer courses and specializations on a wide range of topics. The AI in Healthcare Specialization is a series of courses that cover the fundamentals of AI and its applications in healthcare, including machine learning, natural language processing, and decision support systems.

2. edX: Artificial Intelligence in Healthcare edX is another popular online learning platform that offers courses from leading universities and institutions. The Artificial Intelligence in Healthcare course, offered by Stanford University, provides an introduction to AI and its potential to transform healthcare, covering topics such as machine learning, data mining, and clinical decision support.

3. Udacity: AI for Healthcare Nanodegree. Udacity is an online learning platform that offers project-based courses and nanodegree programs in various technical fields. The AI for Healthcare Nanodegree program is designed for healthcare professionals and data scientists looking to apply AI techniques to healthcare problems, covering topics such as medical image analysis, natural language processing, and reinforcement learning.

10.6.3 Professional Certification Bodies

1. American Board of Artificial Intelligence in Medicine (ABAIM). ABAIM is a professional certification body that offers the Certified Artificial Intelligence in Medicine (CAIM) credential. This certification demonstrates expertise in the application of AI in healthcare and requires passing an examination that covers topics such as machine learning, decision support systems, and ethical considerations.

2. Healthcare Information and Management Systems Society (HIMSS). Certifications HIMSS offers several certification programs relevant to AI in healthcare, including the Certified Associate in Healthcare Information and Management Systems (CAHIMS) and the Certified Professional in Healthcare Information and Management Systems (CPHIMS). These

certifications demonstrate knowledge and skills in healthcare informatics and technology, including AI and machine learning.

3. **American Health Information Management Association (AHIMA).** Certifications AHIMA offers a range of certification programs for healthcare information professionals, including the Registered Health Information Administrator (RHIA) and the Certified Health Data Analyst (CHDA). These certifications cover topics such as data analytics, informatics, and information governance, which are relevant to the application of AI in healthcare.

By pursuing these training and certification programs, you can gain the specialized knowledge and skills needed to effectively apply AI in healthcare settings, advance your career, and contribute to the growing field of healthcare informatics.

10.7 PUTTING IT ALL TOGETHER: CREATING YOUR LEARNING PLAN

Now that you have a comprehensive overview of the resources and opportunities available for continuous learning in AI and healthcare, it is time to create a personalized learning plan. A well-structured learning plan will help you set goals, prioritize your learning activities, and track your progress over time.

10.7.1 Assessing Your Current Knowledge and Skills

Before creating your learning plan, take some time to assess your current knowledge and skills in AI and healthcare. Consider the following questions:

1. What is your current level of understanding of AI concepts and techniques?
2. What experience do you have with applying AI in healthcare settings?
3. What are your strengths and weaknesses in terms of technical skills, domain knowledge, and soft skills?
4. What are the most significant gaps in your knowledge or skills that you would like to address?

By honestly evaluating your current abilities and identifying areas for improvement, you can create a learning plan that targets your specific needs and goals.

10.7.2 Setting Learning Goals and Objectives

Based on your self-assessment, set clear and achievable learning goals and objectives. These should be specific, measurable, and aligned with your professional aspirations. Examples of learning goals might include the following:

1. Develop a deep understanding of machine learning algorithms and their applications in healthcare within the next six months.
2. Gain hands-on experience with implementing an AI-based clinical decision support system within the next year.
3. Obtain a professional certification in healthcare informatics within the next 18 months.

Break down your goals into smaller, actionable objectives that you can work toward on a weekly or monthly basis. For example, if your goal is to develop a deep understanding of machine learning, your objectives might include completing an online course, reading a textbook chapter, and implementing a simple machine learning model.

10.7.3 Crafting a Personalized Learning Roadmap

With your goals and objectives in mind, create a personalized learning roadmap that outlines the specific activities and resources you will use to achieve them. Your roadmap should include a mix of learning activities, such as the following:

1. Formal coursework and training programs
2. Self-directed learning through books, articles, and online resources
3. Hands-on projects and practical applications
4. Attending conferences and workshops
5. Engaging with professional organizations and communities

Assign a timeline to each activity and prioritize them based on their importance and urgency. Be sure to allocate sufficient time for each activity, and balance time with your other professional and personal commitments.

Regularly review and adjust your learning plan based on your progress and changing needs. Celebrate your achievements along the way and use them as motivation to continue your learning journey.

By creating and following a personalized learning plan, you can take control of your professional development and ensure that you are well-equipped to navigate the rapidly evolving field of AI in healthcare. Remember, continuous learning is not a destination, but a lifelong journey of growth and discovery.

As you embark on your continuous learning journey in the field of AI and healthcare, remember that the key to success lies in your commitment, curiosity, and willingness to adapt. The resources and strategies outlined in this chapter provide a solid foundation for your learning, but it is up to you to put them into practice and make the most of the opportunities available.

Here are a few final tips to keep in mind as you navigate your learning journey:

1. Stay curious and open-minded: The field of AI in healthcare is constantly evolving, with new discoveries and innovations emerging all the time. Maintain a sense of curiosity and be open to new ideas and perspectives, even if they challenge your existing knowledge or beliefs.

2. Embrace collaboration and networking: Learning is often a collaborative process, and engaging with others in the field can greatly enhance your knowledge and skills. Seek out opportunities to collaborate on projects, participate in discussions, and learn from the experiences of others.

3. Apply your learning in practice: The true value of your learning lies in your ability to apply it in real-world settings. Look for opportunities to put your knowledge and skills into practice, whether through hands-on projects, case studies, or volunteer work.

4. Share your knowledge with others: As you gain expertise in AI and healthcare, consider sharing your knowledge with others through writing, teaching, or mentoring. Not only will this help solidify your own understanding, but it will also contribute to the growth and advancement of the field as a whole.

5. Engage in lifelong learning: The field of AI in healthcare is always changing. Commit to continuously updating and expanding your knowledge and skills throughout your career.

By following these tips and making continuous learning a priority, you will be well-equipped to navigate the exciting and rapidly evolving field of AI in healthcare. Whether you are a clinician, researcher, or informatics professional, your dedication to learning will enable you to make meaningful contributions to the field and ultimately improve the lives of patients and communities worldwide.

BIBLIOGRAPHY

Abernathy, J. L., Beyer, B., Downes, J. F., & Rapley, E. T. (2020). High-quality information technology and capital investment decisions. *Journal of Information Systems, 34*(3), 1-29.

Agiwal, M., Roy, A., & Saxena, N. (2016). Next generation 5G wireless networks: A comprehensive survey. *IEEE Communications Surveys & Tutorials, 18*(3), 1617-1655.

Ahad, A., Tahir, M., & Yau, K. L. A. (2019). 5G-based smart healthcare network: Architecture, taxonomy, challenges and future research directions. *IEEE Access, 7*, 100747-100762.

Ahmad, M. A., Eckert, C., & Teredesai, A. (2018). Interpretable machine learning in healthcare. In *Proceedings of the 2018 ACM International Conference on Bioinformatics, Computational Biology, and Health Informatics* (pp. 559-560).

Amann, J., Blasimme, A., Vayena, E., Frey, D., & Madai, V. I. (2020). Explainability for artificial intelligence in healthcare: A multidisciplinary perspective. *BMC Medical Informatics and Decision Making, 20*(1), 1-9.

Ardila, D., Kiraly, A. P., Bharadwaj, S., Choi, B., Reicher, J. J., Peng, L., & Shetty, S. (2019). End-to-end lung cancer screening with three-dimensional deep learning on low-dose chest computed tomography. *Nature Medicine, 25*(6), 954-961.

Artetxe, A., Beristain, A., & Graña, M. (2018). Predictive models for hospital readmission risk: A systematic review of methods. *Computer Methods and Programs in Biomedicine, 164*, 49-64.

Baker, J. A., Kornguth, P. J., Lo, J. Y., & Floyd, C. E. (1998). Artificial neural network: Improving the quality of breast biopsy recommendations. *Radiology, 198*(1), 131-135.

Beam, A. L., & Kohane, I. S. (2018). Big data and machine learning in healthcare. *JAMA, 319*(13), 1317-1318.

Berkhin, P. (2006). A survey of clustering data mining techniques. In *Grouping Multidimensional Data* (pp. 25-71). Springer.

Char, D. S., Abràmoff, M. D., & Feudtner, C. (2020). Identifying ethical considerations for machine learning healthcare applications. *The American Journal of Bioethics, 20*(11), 7-17.

Char, D. S., Shah, N. H., & Magnus, D. (2018). Implementing machine learning in healthcare—addressing ethical challenges. *The New England Journal of Medicine, 378*(11), 981-983.

Chen, I. Y., Pierson, E., Rose, S., Joshi, S., Ferryman, K., & Ghassemi, M. (2021). Ethical machine learning in healthcare. *Annual Review of Biomedical Data Science, 4*, 123-144.

Chen, I. Y., Szolovits, P., & Ghassemi, M. (2020). Can AI help reduce disparities in general medical and mental healthcare? *AMA Journal of Ethics, 21*(2), 167-179.

Cho, K., Van Merriënboer, B., Gulcehre, C., Bahdanau, D., Bougares, F., Schwenk, H., & Bengio, Y. (2014). Learning phrase representations using RNN encoder-decoder for statistical machine translation. *arXiv preprint arXiv:1406.1078*.

Choi, E., Biswal, S., Malin, B., Duke, J., Stewart, W. F., & Sun, J. (2017). Generating multi-label discrete patient records using generative adversarial networks. In *Machine Learning for Healthcare Conference* (pp. 286-305).

Choi, E., Schuetz, A., Stewart, W. F., & Sun, J. (2016). Using recurrent neural network models for early detection of heart failure onset. *Journal of the American Medical Informatics Association, 24*(2), 361-370.

Clark, T., Caufield, H., Parker, J. A., Al Manir, S., Amorim, E., Eddy, J., & Munoz-Torres, M. C. (2024). AI-readiness for Biomedical Data: Bridge2AI Recommendations. *bioRxiv*.

Cutillo, C. M., Sharma, K. R., Foschini, L., Kundu, S., Mackintosh, M., & Mandl, K. D. (2020). Machine intelligence in healthcare—perspectives on trustworthiness, explainability, usability, and transparency. *NPJ Digital Medicine, 3*, 47.

Damen, J. A., Hooft, L., Schuit, E., Debray, T. P., Collins, G. S., Tzoulaki, I., & Moons, K. G. (2016). Prediction models for cardiovascular disease risk in the general population: Systematic review. *BMJ, 353*, i2416.

Davis, S. E., Greevy, R. A., Lasko, T. A., Walsh, C. G., & Matheny, M. E. (2020). Comparison of Prediction Model Performance Updating Protocols: Using a Data-Driven Testing Procedure to Guide Updating. *AMIA ... Annual Symposium proceedings. AMIA Symposium, 2019*, 1002–1010.

Demner-Fushman, D., Chapman, W. W., & McDonald, C. J. (2009). What can natural language processing do for clinical decision support? *Journal of Biomedical Informatics, 42*(5), 760-772.

Demner-Fushman, D., Elhadad, N., & Chapman, W. W. (2020). Natural language processing for clinical decision support. In *Clinical Decision Support Systems* (pp. 167-197). Springer.

Denecke, K., & Deng, Y. (2015). Sentiment analysis in medical settings: New opportunities and challenges. *Artificial Intelligence in Medicine, 64*(1), 17-27.

Ding, X., Shang, B., Xie, C. *et al.* Artificial intelligence in the COVID-19 pandemic: balancing benefits and ethical challenges in China's response. *Humanit Soc Sci Commun* **12**, 245 (2025). https://doi.org/10.1057/s41599-025-04564-x

Ding, Y., Sohn, J. H., Kawczynski, M. G., Trivedi, H., Harnish, R., Jenkins, N. W., & Mortensen, E. M. (2018). A deep learning model to predict a diagnosis of Alzheimer disease by using 18F-FDG PET of the brain. *Radiology, 290*(2), 456-464.

Doshi-Velez, F., & Kim, B. (2017). Towards a rigorous science of interpretable machine learning. *arXiv preprint arXiv:1702.08608*.

Dwork, C., & Roth, A. (2014). The algorithmic foundations of differential privacy. *Foundations and Trends in Theoretical Computer Science, 9*(3-4), 211-407.

Esteva, A., Kuprel, B., Novoa, R. A., Ko, J., Swetter, S. M., Blau, H. M., & Thrun, S. (2017). Dermatologist-level classification of skin cancer with deep neural networks. *Nature, 542*(7639), 115-118.

Esteva, A., Robicquet, A., Ramsundar, B., Kuleshov, V., DePristo, M., Chou, K., & Dean, J. (2019). A guide to deep learning in healthcare. *Nature Medicine, 25*(1), 24-29.

Etli, D., Djurovic, A., & Lark, J. (2024). The Future of Personalized Healthcare: AI-Driven Wearables for Real-Time Health Monitoring and Predictive Analytics. *Current Research in Health Sciences, 2*(2), 10-14.

European Parliament and Council of European Union. (2016). Regulation (EU) 2016/679 (General Data Protection Regulation). *Official Journal of the European Union, L119*, 1-88.

Favaretto, M., De Clercq, E., & Elger, B. S. (2019). Big Data and discrimination: Perils, promises and solutions. A systematic review. *Journal of Big Data, 6*(1), 1-27.

Food and Drug Administration (FDA). (2019). Proposed regulatory framework for modifications to artificial intelligence/machine learning (AI/ML)-based software as a medical device (SaMD). Retrieved from *https://www.fda.gov/media/122535/download*

Friedman, J., Hastie, T., & Tibshirani, R. (2001). *The elements of statistical learning*. Springer Series in Statistics.

Frid-Adar, M., Diamant, I., Klang, E., Amitai, M., Goldberger, J., & Greenspan, H. (2018). GAN-based synthetic medical image augmentation for increased CNN performance in liver lesion classification. *Neurocomputing, 321*, 321-331.

Gaffney, A., Woolhandler, S., Cai, C., Bor, D., Himmelstein, J., McCormick, D., & Himmelstein, D. U. (2022). Medical Documentation Burden Among US Office-Based Physicians in 2019: A National Study. *JAMA internal medicine, 182*(5), 564–566. *https://doi.org/10.1001/jamainternmed.2022.0372*

Gawande, M. S., Zade, N., Kumar, P., Gundewar, S., Weerarathna, I. N., & Verma, P. (2025). The role of artificial intelligence in pandemic responses: from epidemiological modeling to vaccine development. *Molecular biomedicine, 6*(1), 1. *https://doi.org/10.1186/s43556-024-00238-3*

Gentry, C. (2009). Fully homomorphic encryption using ideal lattices. In *Proceedings of the forty-first annual ACM symposium on Theory of computing* (pp. 169-178).

Gianfrancesco, M. A., Tamang, S., Yazdany, J., & Schmajuk, G. (2018). Potential biases in machine learning algorithms using electronic health record data. *JAMA Internal Medicine, 178*(11), 1544-1547.

Gilpin, L. H., Bau, D., Yuan, B. Z., Bajwa, A., Specter, M., & Kagal, L. (2018). Explaining explanations: An overview of interpretability of machine learning. In *2018 IEEE 5th International Conference on Data Science and Advanced Analytics (DSAA)* (pp. 80-89).

Ginsburg, G. S., & Phillips, K. A. (2018). Precision medicine: From science to value. *Health Affairs, 37*(5), 694-701.

Glorot, X., Bordes, A., & Bengio, Y. (2011). Deep sparse rectifier neural networks. In *Proceedings of the Fourteenth International Conference on Artificial Intelligence and Statistics* (pp. 315-323).

Goldstein, B. A., Navar, A. M., Pencina, M. J., & Ioannidis, J. P. A. (2017). Opportunities and challenges in developing risk prediction models with electronic health records data: A systematic review. *Journal of the American Medical Informatics Association, 24*(1), 198-208.

Goodfellow, I., Bengio, Y., & Courville, A. (2016). *Deep learning*. MIT Press.

Goodfellow, I., Pouget-Abadie, J., Mirza, M., Xu, B., Warde-Farley, D., Ozair, S., & Bengio, Y. (2014). Generative adversarial nets. *Advances in Neural Information Processing Systems, 27*.

Greaves, F., Ramirez-Cano, D., Millett, C., Darzi, A., & Donaldson, L. (2013). Use of sentiment analysis for capturing patient experience from free-text comments posted online. *Journal of Medical Internet Research, 15*(11), e239.

Gräßer, F., Kallumadi, S., Malberg, H., & Zaunseder, S. (2018). Aspect-based sentiment analysis of drug reviews applying cross-domain and cross-data learning. In *Proceedings of the 2018 International Conference on Digital Health* (pp. 121-125).

Guidotti, R., Monreale, A., Ruggieri, S., Turini, F., Giannotti, F., & Pedreschi, D. (2018). A survey of methods for explaining black box models. *ACM Computing Surveys, 51*(5), 1-42.

Gulshan, V., Peng, L., Coram, M., Stumpe, M. C., Wu, D., Narayanaswamy, A., & Webster, D. R. (2016). Development and validation of a deep learning algorithm for detection of diabetic retinopathy in retinal fundus photographs. *JAMA, 316*(22), 2402-2410.

Guntuku, S. C., Yaden, D. B., Kern, M. L., Ungar, L. H., & Eichstaedt, J. C. (2017). Detecting depression and mental illness on social media: An integrative review. *Current Opinion in Behavioral Sciences, 18*, 43-49.

Hahsler, M., & Karpienko, R. (2017). Visualizing association rules in hierarchical groups. *Journal of Business Economics, 87*(3), 317-335.

Han, J., Pei, J., & Kamber, M. (2022). *Data mining: Concepts and techniques*. Morgan Kaufmann.

Hastie, T., Tibshirani, R., & Friedman, J. (2009). *The elements of statistical learning: Data mining, inference, and prediction.* Springer Science & Business Media.

Hersh, W. R., Totten, A. M., Eden, K. B., Devine, B., Gorman, P., Kassakian, S. Z., & McDonagh, M. S. (2015). Outcomes from health information exchange: Systematic review and future research needs. *JMIR Medical Informatics, 3*(4), e39.

Heyndrickx, W., Mervin, L., Morawietz, T., Sturm, N., Friedrich, L., Zalewski, A., Pentina, A., Humbeck, L., Oldenhof, M., Niwayama, R., Schmidtke, P., Fechner, N., Simm, J., Arany, A., Drizard, N., Jabal, R., Afanasyeva, A., Loeb, R., Verma, S., Harnqvist, S., & Ceulemans, H. (2024). MELLODDY: Cross-pharma Federated Learning at Unprecedented Scale Unlocks Benefits in QSAR without Compromising Proprietary Information. *Journal of Chemical Information and Modeling, 64*(7), 2331–2344. https://doi.org/10.1021/acs.jcim.3c00799

Hinton, G. E., & Sejnowski, T. J. (Eds.). (1999). *Unsupervised learning: Foundations of neural computation.* MIT Press.

Hochreiter, S., & Schmidhuber, J. (1997). Long short-term memory. *Neural Computation, 9*(8), 1735-1780.

Hripcsak, G., Albers, D. J., & Perotte, A. (2015). Parameterizing time in electronic health record studies. *Journal of the American Medical Informatics Association, 22*(4), 794-804.

Huang, Z., Dong, W., Ji, L., Gan, C., Lu, X., & Duan, H. (2014). Discovery of clinical pathway patterns from event logs using probabilistic topic models. *Journal of Biomedical Informatics, 47*, 39-57.

Jagannatha, A. N., & Yu, H. (2016). Bidirectional RNN for medical event detection in electronic health records. In *Proceedings of the Conference of the North American Chapter of the Association for Computational Linguistics.*

James, G., Witten, D., Hastie, T., & Tibshirani, R. (2021). *An introduction to statistical learning.* Springer.

Jiang, F., Jiang, Y., Zhi, H., Dong, Y., Li, H., Ma, S., & Wang, Y. (2017). Artificial intelligence in healthcare: Past, present and future. *Stroke and Vascular Neurology, 2*(4), 230-243.

Jing, B., Xie, P., & Xing, E. (2017). On the automatic generation of medical imaging reports. *arXiv preprint arXiv:1711.08195*.

Kamnitsas, K., Ledig, C., Newcombe, V. F., Simpson, J. P., Kane, A. D., Menon, D. K., & Glocker, B. (2017). Efficient multi-scale 3D CNN with fully connected CRF for accurate brain lesion segmentation. *Medical Image Analysis, 36*, 61-78.

Kang, J., Schwartz, R., Flickinger, J., & Beriwal, S. (2021). Machine learning approaches for predicting radiation therapy outcomes: A clinician's perspective. *International Journal of Radiation Oncology, Biology, Physics, 109*(2), 303-313.

Kaur, R., & Ginige, J. A. (2018). Comparative analysis of algorithmic approaches for auto-coding with ICD-10-AM and ACHI. *Health Information Science and Systems, 6*(1), 1-13.

Kindig, D., & Stoddart, G. (2003). What is population health? *American Journal of Public Health, 93*(3), 380-383.

Koleck, T. A., Dreisbach, C., Bourne, P. E., & Bakken, S. (2019). Natural language processing of symptoms documented in free-text narratives of electronic health records: A systematic review. *Journal of the American Medical Informatics Association, 26*(4), 364-379.

Krizhevsky, A., Sutskever, I., & Hinton, G. E. (2012). Imagenet classification with deep convolutional neural networks. *Advances in Neural Information Processing Systems, 25*, 1097-1105.

Kuhn, M., & Johnson, K. (2019). Feature Engineering and Selection: A Practical Approach for Predictive Models (1st ed.). Chapman and Hall/CRC. *https://doi.org/10.1201/9781315108230*

Kulikowski, C. A., Shortliffe, E. H., Currie, L. M., Elkin, P. L., Hunter, L. E., Johnson, T. R., & Williamson, J. J. (2012). AMIA Board white paper: Definition of biomedical informatics and specification of core competencies for graduate education in the discipline. *Journal of the American Medical Informatics Association, 19*(6), 931-938.

Kutafina, E., Bechtold, I., Kabino, K., & Jonas, S. M. (2019). Recursive neural networks in hospital demand forecasting. *Healthcare Management Science, 22*(1), 144-156.

Lalmuanawma, S., Hussain, J., & Chhakchhuak, L. (2020). Applications of machine learning and artificial intelligence for Covid-19 (SARS-CoV-2) pandemic: A review. *Chaos, Solitons & Fractals, 139*, 110059.

LeCun, Y., Bottou, L., Bengio, Y., & Haffner, P. (1998). Gradient-based learning applied to document recognition. *Proceedings of the IEEE, 86*(11), 2278-2324.

Ledley, R. S., & Lusted, L. B. (1959). Reasoning foundations of medical diagnosis. *Science, 130*(3366), 9-21.

Levin, S., Toerper, M., Hamrock, E., Hinson, J. S., Barnes, S., Gardner, H., Linton, B., Kirsch, T., & Kelen, G. (2018). Machine-learning-based electronic triage more accurately differentiates patients with respect to clinical outcomes compared with the emergency severity index. *Annals of Emergency Medicine, 71*(5), 565-574.

Li, Z., He, Y., Keel, S., Meng, W., Chang, R. T., & He, M. (2018). Efficacy of a deep learning system for detecting glaucomatous optic neuropathy based on color fundus photographs. *Ophthalmology, 125*(8), 1199-1206.

Liao, M., Li, Y., Kianifard, F., Obi, E., & Arcona, S. (2016). Cluster analysis and its application to healthcare claims data: A study of end-stage renal disease patients who initiated hemodialysis. *BMC Nephrology, 17*(1), 1-14.

Lipton, Z. C., Berkowitz, J., & Elkan, C. (2015). A critical review of recurrent neural networks for sequence learning. *arXiv preprint arXiv:1506.00019*.

Litjens, G., Kooi, T., Bejnordi, B. E., Setio, A. A. A., Ciompi, F., Ghafoorian, M., & Sánchez, C. I. (2017). A survey on deep learning in medical image analysis. *Medical Image Analysis, 42*, 60-88.

Litjens, G., Sánchez, C. I., Timofeeva, N., Hermsen, M., Nagtegaal, I., Kovacs, I., & Van Der Laak, J. (2016). Deep learning as a tool for increased accuracy and efficiency of histopathological diagnosis. *Scientific Reports, 6*(1), 1-11.

Liu, B. (2012). *Sentiment analysis and opinion mining*. Synthesis Lectures on Human Language Technologies.

Liu, X., Faes, L., Kale, A. U., Wagner, S. K., Fu, D. J., Bruynseels, A., & Denniston, A. K. (2019). A comparison of deep learning performance against health-care professionals in detecting diseases from medical imaging: A systematic review and meta-analysis. *The Lancet Digital Health, 1*(6), e271-e297.

Luo, Y., Uzuner, Ö., & Szolovits, P. (2017). Bridging semantics and syntax with graph algorithms—state-of-the-art of extracting biomedical relations. *Briefings in Bioinformatics, 18*(1), 160-178.

Lundberg, S. M., & Lee, S. I. (2017). A unified approach to interpreting model predictions. In *Proceedings of the 31st International Conference on Neural Information Processing Systems* (pp. 4768-4777).

Makino, M., Yoshimoto, R., Ono, M., Itoko, T., Katsuki, T., Koseki, A., & Kudo, M. (2019). Artificial intelligence predicts the progression of diabetic kidney disease using big data machine learning. *Scientific Reports, 9*(1), 1-9.

Manary, M. P., Boulding, W., Staelin, R., & Glickman, S. W. (2013). The patient experience and health outcomes. *New England Journal of Medicine, 368*(3), 201-203.

McKinney, S. M., Sieniek, M., Godbole, V., Godwin, J., Antropova, N., Ashrafian, H., & Shetty, S. (2020). International evaluation of an AI system for breast cancer screening. *Nature, 577*(7788), 89-94.

Medhat, W., Hassan, A., & Korashy, H. (2014). Sentiment analysis algorithms and applications: A survey. *Ain Shams Engineering Journal, 5*(4), 1093-1113.

Mehrabi, N., Morstatter, F., Saxena, N., Lerman, K., & Galstyan, A. (2021). A survey on bias and fairness in machine learning. *ACM Computing Surveys, 54*(6), 1-35.

Meystre, S. M., Savova, G. K., Kipper-Schuler, K. C., & Hurdle, J. F. (2008). Extracting information from textual documents in the electronic health record: A review of recent research. *Yearbook of Medical Informatics, 17*(01), 128-144.

Miotto, R., Wang, F., Wang, S., Jiang, X., & Dudley, J. T. (2018). Deep learning for healthcare: Review, opportunities, and challenges. *Briefings in Bioinformatics, 19*(6), 1236-1246.

Mitchell, T. M. (1997). *Machine learning*. McGraw-Hill.

Mnih, V., Kavukcuoglu, K., Silver, D., Rusu, A. A., Veness, J., Bellemare, M. G., & Hassabis, D. (2015). Human-level control through deep reinforcement learning. *Nature, 518*(7540), 529-533.

Muehlematter, U. J., Daniore, P., & Vokinger, K. N. (2021). Approval of artificial intelligence and machine learning-based medical devices in the USA and Europe (2015–20): A comparative analysis. *The Lancet Digital Health, 3*(3), e195-e203.

Murdoch, T. B., & Detsky, A. S. (2013). The inevitable application of big data to healthcare. *JAMA, 309*(13), 1351-1352.

Murphy, S. N., Weber, G., Mendis, M., Gainer, V., Chueh, H. C., Churchill, S., & Kohane, I. (2010). Serving the enterprise and beyond with informatics for integrating biology and the bedside (i2b2). *Journal of the American Medical Informatics Association, 17*(2), 124-130.

Nair, V., & Hinton, G. E. (2010). Rectified linear units improve restricted Boltzmann machines. In *Proceedings of the 27th International Conference on Machine Learning (ICML-10)* (pp. 807-814).

Nguyen, D. C., Ding, M., Pathirana, P. N., & Seneviratne, A. (2021). Federated learning for internet of things: A comprehensive survey. *IEEE Communications Surveys & Tutorials, 23*(3), 1622-1658.

Obermeyer, Z., & Emanuel, E. J. (2016). Predicting the future—big data, machine learning, and clinical medicine. *The New England Journal of Medicine, 375*(13), 1216.

Office of the National Coordinator for Health Information Technology. (2021). Hospital Progress to Meaningful Use by Size, Type, and Urban/Rural Location. Health IT Quick-Stat #56. *https://www.healthit.gov/data/quickstats/hospital-progress-meaningful-use-size-type-and-urbanrural-location*

Olah, C., Mordvintsev, A., & Schubert, L. (2017). Feature visualization. *Distill, 2*(11), e7.

Ordonez, C., Ezquerra, N., & Santana, C. A. (2006). Constraining and summarizing association rules in medical data. *Knowledge and Information Systems, 9*(3), 259-283.

Pang, J., Huang, Y., Xie, Z., Li, J., & Cai, Z. (2021). Collaborative city digital twin for the COVID-19 pandemic: A federated learning solution. *Tsinghua Science and Technology, 26*(5), 759-771.

Parikh, R. B., Manz, C., Chivers, C., Regli, S. H., Braun, J., Draugelis, M. E., & Bates, D. W. (2019). Machine learning approaches to predict 6-month mortality among patients with cancer. *JAMA Network Open, 2*(10), e1915997.

Payrovnaziri, S. N., Chen, Z., Rengifo-Moreno, P., Miller, T., Bian, J., Chen, J. H., & He, Z. (2020). Explainable artificial intelligence models using real-world electronic health record data: A systematic scoping review. *Journal of the American Medical Informatics Association, 27*(7), 1173-1185.

Perou, C. M., Sørlie, T., Eisen, M. B., Van De Rijn, M., Jeffrey, S. S., Rees, C. A., & Botstein, D. (2000). Molecular portraits of human breast tumours. *Nature, 406*(6797), 747-752.

Pirandola, S., Andersen, U. L., Banchi, L., Berta, M., Bunandar, D., Colbeck, R., & Wallden, P. (2020). Advances in quantum cryptography. *Advances in Optics and Photonics, 12*(4), 1012-1236.

Pivovarov, R., & Elhadad, N. (2015). Automated methods for the summarization of electronic health records. *Journal of the American Medical Informatics Association, 22*(5), 938-947.

Porter, M. E., & Teisberg, E. O. (2006). *Redefining healthcare: Creating value-based competition on results*. Harvard Business Press.

Qiu, H., Qiu, M., Liu, M., & Memmi, G. (2020). Secure health data sharing for medical cyber-physical systems for the healthcare 4.0. *IEEE Journal of Biomedical and Health Informatics, 24*(9), 2499-2505.

Rahman, M. A., Hossain, M. S., Alrajeh, N. A., & Guizani, N. (2020). B5G and explainable deep learning assisted healthcare vertical at the edge: COVID-19 perspective. *IEEE Network, 34*(4), 98-105.

Rajkomar, A., Oren, E., Chen, K., Dai, A. M., Hajaj, N., Hardt, M., & Dean, J. (2018). Scalable and accurate deep learning with electronic health records. *NPJ Digital Medicine, 1*(1), 1-10.

Ramachandram, D., & Taylor, G. W. (2017). Deep multimodal learning: A survey on recent advances and trends. *IEEE Signal Processing Magazine, 34*(6), 96-108.

Ramesh, B. P., Houston, T. K., Brandt, C., Fang, H., & Yu, H. (2013). Improving patients' electronic health record comprehension with NoteAid. In *MedInfo* (pp. 714-718).

Raghu, M., Zhang, C., Kleinberg, J., & Bengio, S. (2019). Transfusion: Understanding transfer learning for medical imaging. *arXiv preprint arXiv:1902.07208*.

Ribeiro, M. T., Singh, S., & Guestrin, C. (2016). "Why should I trust you?" Explaining the predictions of any classifier. In *Proceedings of the 22nd ACM SIGKDD International Conference on Knowledge Discovery and Data Mining* (pp. 1135-1144).

Rieke, N., Hancox, J., Li, W., Milletari, F., Roth, H. R., Albarqouni, S., & Cardoso, M. J. (2020). The future of digital health with federated learning. *NPJ Digital Medicine, 3*(1), 119.

Ronneberger, O., Fischer, P., & Brox, T. (2015). U-net: Convolutional networks for biomedical image segmentation. In *International Conference on Medical Image Computing and Computer-Assisted Intervention* (pp. 234-241).

Rosenbloom, S. T., Denny, J. C., Xu, H., Lorenzi, N., Stead, W. W., & Johnson, K. B. (2011). Data from clinical notes: A perspective on the tension between structure and flexible documentation. *Journal of the American Medical Informatics Association, 18*(2), 181-186.

Rumelhart, D. E., Hinton, G. E., & Williams, R. J. (1986). Learning representations by back-propagating errors. *Nature, 323*(6088), 533-536.

Russell, S., & Norvig, P. (2021). *Artificial intelligence: A modern approach* (4th ed.). Pearson.

Satyanarayanan, M. (2017). The emergence of edge computing. *Computer, 50*(1), 30-39.

Schork, N. J. (2019). Artificial intelligence and personalized medicine. *Cancer Treatment and Research, 178*, 265-283.

Sendak, M., Gao, M., Nichols, M., Lin, A., & Balu, S. (2019). Machine learning in healthcare: A critical appraisal of challenges and opportunities. *EGEMS (Washington, DC), 7*(1), 1.

Sheikhalishahi, S., Miotto, R., Dudley, J. T., Lavelli, A., Rinaldi, F., & Osmani, V. (2019). Natural language processing of clinical notes on chronic diseases: systematic review. *JMIR Medical Informatics*, 7(2), e12239.

Shen, D., Wu, G., & Suk, H. I. (2017). Deep learning in medical image analysis. *Annual Review of Biomedical Engineering*, 19, 221-248.

Shen, Y., Zheng, K., Guo, Y., Chen, Y., Zhao, X., Huang, L., & Feng, M. (2019). An interpretable classifier for high-resolution breast cancer screening images utilizing weakly supervised localization. *Medical Image Analysis*, 68, 101908.

Shi, W., Cao, J., Zhang, Q., Li, Y., & Xu, L. (2016). Edge computing: Vision and challenges. *IEEE Internet of Things Journal*, 3(5), 637-646.

Shickel, B., Tighe, P. J., Bihorac, A., & Rashidi, P. (2018). Deep EHR: A survey of recent advances in deep learning techniques for electronic health record (EHR) analysis. *IEEE Journal of Biomedical and Health Informatics*, 22(5), 1589-1604.

Shortliffe, E. H. (1976). *Computer-based medical consultations: MYCIN* (Vol. 2). Elsevier.

Singhal, S., Kayyali, B., Levin, R., & Greenberg, Z. (2020). The potential for artificial intelligence in healthcare. *McKinsey & Company*.

Sinnenberg, L., Buttenheim, A. M., Padrez, K., Mancheno, C., Ungar, L., & Merchant, R. M. (2017). Twitter as a tool for health research: A systematic review. *American Journal of Public Health*, 107(1), e1-e8.

Sinsky, C., Colligan, L., Li, L., Prgomet, M., Reynolds, S., Goeders, L., & Blike, G. (2016). Allocation of physician time in ambulatory practice: A time and motion study in 4 specialties. *Annals of Internal Medicine*, 165(11), 753-760.

Soni, J., Ansari, U., Sharma, D., & Soni, S. (2011). Predictive data mining for medical diagnosis: An overview of heart disease prediction. *International Journal of Computer Applications*, 17(8), 43-48.

Steyerberg, E. W. (2019). *Clinical prediction models*. Springer International Publishing.

Sun, W., Rumshisky, A., & Uzuner, O. (2013). Evaluating temporal relations in clinical text: 2012 i2b2 Challenge. *Journal of the American Medical Informatics Association*, 20(5), 806-813.

Sutton, R. S., & Barto, A. G. (2018). *Reinforcement learning: An introduction.* MIT Press.

Sutton, R. T., Pincock, D., Baumgart, D. C., Sadowski, D. C., Fedorak, R. N., & Kroeker, K. I. (2020). An overview of clinical decision support systems: Benefits, risks, and strategies for success. *NPJ Digital Medicine, 3*(1), 1-10.

Syeda, H. B., Syed, M., Sexton, K. W., Syed, S., Begum, S., Syed, F., & Yu, F. (2021). Role of machine learning techniques to tackle the COVID-19 crisis: Systematic review. *JMIR Medical Informatics, 9*(1), e23811.

Tan, P. N., Steinbach, M., & Kumar, V. (2016). *Introduction to data mining.* Pearson Education India.

Tang, D., Qin, B., & Liu, T. (2016). Aspect level sentiment classification with deep memory network. *arXiv preprint arXiv:1605.08900.*

Tatonetti, N. P., Ye, P. P., Daneshjou, R., & Altman, R. B. (2012). Data-driven prediction of drug effects and interactions. *Science Translational Medicine, 4*(125), 125ra31-125ra31.

Tomašev, N., Glorot, X., Rae, J. W., Zielinski, M., Askham, H., Saraiva, A., & Protsyuk, I. (2019). A clinically applicable approach to continuous prediction of future acute kidney injury. *Nature, 572*(7767), 116-119.

Tonekaboni, S., Joshi, S., McCradden, M. D., & Goldenberg, A. (2019). What clinicians want: Contextualizing explainable machine learning for clinical end use. In *Machine Learning for Healthcare Conference* (pp. 359-380).

Topol, E. J. (2019). High-performance medicine: The convergence of human and artificial intelligence. *Nature Medicine, 25*(1), 44-56.

Uzuner, Ö., South, B. R., Shen, S., & DuVall, S. L. (2011). 2010 i2b2/VA challenge on concepts, assertions, and relations in clinical text. *Journal of the American Medical Informatics Association, 18*(5), 552-556.

Vaswani, A., Shazeer, N., Parmar, N., Uszkoreit, J., Jones, L., Gomez, A. N., & Polosukhin, I. (2017). Attention is all you need. In *Advances in Neural Information Processing Systems* (pp. 5998-6008).

Vuik, S. I., Mayer, E. K., & Darzi, A. (2016). Patient segmentation analysis offers significant benefits for integrated care and support. *Health Affairs, 35*(5), 769-775.

Wang, Y., Wang, L., Rastegar-Mojarad, M., Moon, S., Shen, F., Afzal, N., & Liu, H. (2018). Clinical information extraction applications: A literature review. *Journal of Biomedical Informatics, 77*, 34-49.

Wiens, J., Saria, S., Sendak, M., Ghassemi, M., Liu, V. X., Doshi-Velez, F., & Goldenberg, A. (2019). Do no harm: A roadmap for responsible machine learning for healthcare. *Nature Medicine, 25*(9), 1337-1340.

Willemink, M. J., Koszek, W. A., Hardell, C., Wu, J., Fleischmann, D., Harvey, H., Folio, L. R., Summers, R. M., Rubin, D. L., & Lungren, M. P. (2020). Preparing medical imaging data for machine learning. Radiology, 295(1), 4-15.

Wolterink, J. M., Dinkla, A. M., Savenije, M. H., Seevinck, P. R., van den Berg, C. A., & Išgum, I. (2017). Deep MR to CT synthesis using unpaired data. In *International Workshop on Simulation and Synthesis in Medical Imaging* (pp. 14-23).

Wosik, J., Fudim, M., Cameron, B., Gellad, Z. F., Cho, A., Phinney, D., & Tcheng, J. (2020). Telehealth transformation: COVID-19 and the rise of virtual care. *Journal of the American Medical Informatics Association, 27*(6), 957-962.

Wu, S., Roberts, K., Datta, S., Du, J., Ji, Z., Si, Y., & Xu, H. (2020). Deep learning in clinical natural language processing: A methodical review. *Journal of the American Medical Informatics Association, 27*(3), 457-470.

Wynants, L., Van Calster, B., Collins, G. S., Riley, R. D., Heinze, G., Schuit, E., ... & Van Smeden, M. (2020). Prediction models for diagnosis and prognosis of COVID-19 infection: Systematic review and critical appraisal. *The Lancet Digital Health, 2*(6), e411–e426. https://doi.org/10.1016/S2589-7500(20)30217-X

Xiao, C., Choi, E., & Sun, J. (2018). Opportunities and challenges in developing deep learning models using electronic health records data: A systematic review. *Journal of the American Medical Informatics Association, 25*(10), 1419-1428.

Xu, D., & Tian, Y. (2015). A comprehensive survey of clustering algorithms. *Annals of Data Science, 2*(2), 165-193.

Xu, R., & Wunsch, D. (2005). Survey of clustering algorithms. *IEEE Transactions on Neural Networks, 16*(3), 645-678.

Yang, Q., Liu, Y., Chen, T., & Tong, Y. (2019). Federated machine learning: Concept and applications. *ACM Transactions on Intelligent Systems and Technology, 10*(2), 1-19.

Zhang, B. H., Lemoine, B., & Mitchell, M. (2018). Mitigating unwanted biases with adversarial learning. In *Proceedings of the 2018 AAAI/ACM Conference on AI, Ethics, and Society* (pp. 335-340).

Zhang, L., Wang, S., & Liu, B. (2018). Deep learning for sentiment analysis: A survey. *Wiley Interdisciplinary Reviews: Data Mining and Knowledge Discovery, 8*(4), e1253.

Zhang, Z., Zhao, Y., Liao, X., Shi, W., Li, K., Zou, Q., & Peng, S. (2019). Deep learning in omics: A survey and guideline. *Briefings in Functional Genomics, 18*(1), 41-57.

Zitnik, M., Agrawal, M., & Leskovec, J. (2018). Modeling polypharmacy side effects with graph convolutional networks. *Bioinformatics, 34*(13), i457-i466.

INDEX

A

AI in healthcare
  actionable insights in real time, 19
  administrative demands, 23
  AI-assisted diagnostics, 13
  AI-driven treatment optimization, 15–16
  billing systems, 24
  in cardiology, 16
  CDSSs, 19
  challenges and future directions, 14
  chronic illnesses, 28
  clinical documentation, automation of, 23
  clinician acceptance and workflow integration, 20–21
  continuous learning and feedback loops, 21
  continuous learning health systems, 17
  early detection and disease screening, 12
  edge computing
    Alzheimer's disease, early detection of, 439
    decentralized clinical trials and research, 438
    federated learning framework, 454
    5G networks and telemedicine, 455
    heightened privacy and security, 438
    real-time patient monitoring and analysis, 438
    telemedicine and remote care, 438
  epidemiological modeling, 27
  ethical and privacy considerations, 17, 29–30
  evidence-based recommendations, 20
  explainable AI (XAI), 463
  FDA's guidance on AI/ML-based software, 455–458
  federated learning, 460–463
  future directions and challenges, 21–23, 25–27
  history of, 8–10
  improved diagnostic accuracy, 11
  interpretable AI models, 463–464
  lifestyle and behavioral factors, 28
  medical imaging, advancements in, 12
  in mental health, 16
  multimodal data integration, 20
  multimodal diagnostic insights, 13
  multimodal learning and sensor fusion, 465–467

multi-omics integration, 17
multiple data stream integration, 16
in oncology, 16
personalized medicine, 14–15
population health management, 27, 30
precision medicine and personalized care
  challenges and considerations, 434–435
  population health and disease prevention, 434
predictive resource allocation, 24–25
quality assurance and compliance, 25
quantum computing, 436–437
real-time personalization, 17
regulatory bodies and real-world utility, 13–14
reimbursement and incentive structures, 459–460
resource allocation and health system planning, 29
scheduling and appointment management, 24
social determinants of health (SDOH), 28
Alzheimer's disease, early detection of, 439
Apriori Algorithm, 160
Association rule mining (ARM)
  advanced techniques, 180
  algorithms, 160–161
  best practices, 180–181
  challenges, 180
  healthcare applications, 161
  interpretation, 179
  practical implementation of, 161–179
  strengths, 160
Autoregressive modeling, 72

B

Backpropagation, 57
Bag of Words (BoW) approach, 33, 34, 37

C

CHARM algorithm, 161
Clinical Decision Support Systems (CDSSs), 8, 19, 34
  AI techniques in
    clinician trust, 276–277
    data loading and preprocessing, 271
    deep learning architectures, 253
    drug-drug interactions (DDIs), 263–265
    hybrid AI approaches, 254
    hybrid model, 274–276
    knowledge-based rules, 272–274
    knowledge representation and reasoning, 253
    patient risk stratification, 270
    practical considerations, 254–255
    random forest classifier, 272
  chronic diseases
    complexity and heterogeneity, 394
    data integration, 395
    machine learning models, 395
    personalized treatment recommendation system, 396
    treatment recommendations, 395
  continuous improvement and learning, 432
  data quality and integration, 431
  effectiveness of, evaluating, 277–279
  emergency departments (EDs), 418–431
  evaluation
    clinical performance metrics, 277
    economic impact, 278

user acceptance and satisfaction, 278
future perspectives, 432
integration of, 252
model transparency and interpretability, 432
multidisciplinary collaboration, 431
post-surgical complications, 405–418
primary goal of, 251–252
rare diseases
   challenges, 384, 385
   cutting-edge machine learning algorithms, 385, 386
   highlights, 393–394
   knowledge graphs, 384–385
   NLP techniques, 384
   retrospective evaluation, 386
user-centered design, 431
Clinical informatics
  CDSSs
   AI techniques in, 253–277
   effectiveness of, evaluating, 277–279
   evaluation, 277–278
   integration of, 252
   primary goal of, 251–252
  challenges and considerations, 2
  clinical assistant to innovators, 4
  computer vision, 5
  data governance and regulatory compliance, 279
   data protection and anonymization, 281–287
   healthcare organizations, considerations for, 280–281
  EHRs (*see* Electronic health Records (EHRs))
  ethical and philosophical reckonings, 7
  implications, 5–6
  key components, 8
  machine learning and deep learning models, 4
  natural language processing (NLP), 5
  patients data and knowledge, 3–4
  real-world implications, 2
  robotic systems, 5
  scope of, 237
  systems-level impact, 6–7
  telemedicine
   barriers, 248–251
   benefits of, 245
   in COVID-19 pandemic, 246
   remote patient monitoring (RPM), 246–248
   service offerings, 245
Clustering
  best practices, 233
  care pathway analysis, 183
  challenges and limitations, 183, 232–233
  clinical applications, 184, 231–232
  density-based clustering, 182, 185
  disease subtyping, 183
  future directions, 233–234
  goals, 182
  hierarchical clustering, 182, 184
  model-based clustering, 182, 185
  partitional clustering, 182, 184
  patient segmentation, 183
  for patient segmentation, 185–230
Continuous learning, 469–470
  in CDSSs, 254
  combination of strategies, 470–471
  conferences and organizations, 473–475
  and feedback loops, 21
  health systems, 17

information sources, curation of, 471–472
reading recommendations and reference materials
  journals and publications, 472–473
  online articles and blogs, 473
structured learning plan, 477–479
training and certification programs, 475–477
Convolutional Neural Networks (CNNs), 57–58
  in clinical text mining tasks, 117
  medical image analysis, 255–259
  multimodal, 466
  in sentiment analysis tasks, 119

D

Data anonymization, 282
Data mining, 159
Deep learning, 55
  challenges, 61
  clinical applications in healthcare, 59–61
  clinical NLP, 153
  Convolutional Neural Networks (CNNs), 57–58
  future directions, 69–70
  Generative Adversarial Networks (GANs), 58–59
  Recurrent Neural Networks (RNNs), 58
  revolution of, 9
Deep neural networks, 57, 59, 61, 78, 368, 369
Drug discovery and development
  drug repurposing process, 313
  load and preprocess clinical data, 314–315
  use of deep learning models, 313
  use of machine learning algorithms, 313
  Variationl Autoencoder (VAE) approach, 315–321

E

Eclat (Equivalence Class Transformation), 161
Edge computing
  Alzheimer's disease, early detection of, 439
  decentralized clinical trials and research, 438
  federated learning framework, 454
  5G networks and telemedicine, 455
  heightened privacy and security, 438
  real-time patient monitoring and analysis, 438
  telemedicine and remote care, 438
Electronic Health Records (EHRs), 1, 2, 8
  advantages to healthcare, 239
  AI in EHR optimization
    challenges and considerations, 240–241
    future of, 241
    intelligent and adaptive user interfaces, 240
    machine learning algorithms, 240
    NLP algorithms, 240
    problem associated with, 240
  challenges, 239
    data quality and standardization, 241–242
    interoperability of EHR systems, 242
    privacy and security concerns, 242
  clinical decision support tools, 239
  evolution of, 238–244

future perspectives, 242–244
history of, 9
HITECH Act, 238

F

Frequent Pattern (FP) Growth, 161

G

Generative Adversarial Networks (GANs), 58–59

H

Health Information Exchange (HIE), 8

K

Knowledge graphs approach, 253

M

Machine learning
  algorithms in
    decision trees, 42
    linear regression, 41
    logistic regression, 41
    random forests, 42
    support vector machines, 42
  CDSSs, 385, 386, 395
  clinical applications, 48–54
  in clinical informatics, 4
  in drug discovery and development, 313
  EHRs, 240
  rare disease identification, 302
  reinforcement learning, 47
  supervised learning
    aim, 38
    data collection and labeling, 39
    data preparation, 40
    feature engineering, 40
    fundamental steps, 39
    healthcare applications, 42–44
    model deployment, 41
    model evaluation, 40–41
    model training and validation, 40
  unsupervised learning
    autoencoders, 45
    healthcare applications, 45–46
    Hierarchical Clustering, 44–45
    K-means Clustering, 44
    Principal Component Analysis (PCA), 45
Machine Learning Ledger Orchestration for Drug Discovery (MELLODDY) platform, 462
Medical imaging
  field of digital pathology, 293
  pathology reports and laboratory test results, 296–302
  pneumonia detection prediction, 293–296
  in radiology, 292
  types of, 292
Multilevel association rule mining, 180
Multimodal transformers
  sentiment analysis, 118–120
  text mining and information extraction
    clinical text mining tasks, 86–93
    concept normalization, 93–98
    NLP pipelines, 109–118
    temporal information extraction, 98–109

N

Named entity recognition (NER), 86–93
Natural language processing (NLP), 5, 32

in clinical documentation
  challenges, 154–155
  clinical decision support, 142–153
  deep learning models, 153
  evaluation considerations, 155–156
  future of, 156–157
  hybrid approaches, 153–154
  medical coding, 121–125
  NER, 126–134
  summarization, 134–142
clinical text mining, challenges in, 117
sentiment analysis, 118, 119
transformer architectures
  Alibi method, 84
  decoder-only transformer, 72–73
  encoder and decoder, 71
  encoder-only, 72
  layer normalization technique, 78–79
  multimodal transformers, 85–120
  positional encoding, 79–81
  position-wise feedforward layer, 75–77
  residual connections, 77–78
  RoPE, 82–84
  self-attention mechanism, 74–75
  Vision Transformer (ViT), 84
word tokenization, 36
Negative association rule mining, 180
Neural networks, 55
  aggregation and activation step, 56–57
  PyTorch
    batch normalization, 62
    data preparation and model initialization, 63–64
    dropout technique, 62
    He (Kaiming) initialization, 62
    Loss Function, 64
    model inference and interpretability, 67–68
    model.train() and model.eval(), 65–67
    optimizer, and evaluation metrics, 64–65
    sensitivity and specificity, 65
    SHAP framework, 69
  types of layers, 56

P

Positional encoding
  learned positional embeddings, 80
  sinusoidal positional embeddings, 80–81
Predictive analytics, 159
Predictive modeling
  adverse event prediction, 236
  challenges and limitations, 236
  goals, 234
  readmission prediction, 235–236
  risk prediction, 235
  treatment response prediction, 236
  types of, 234–235

Q

Quantitative association rule mining, 180
Quantum computing, 436–437

R

Rare disease identification
  aiding diagnosis, 303
  from EHR data, 303–307
  by machine learning algorithms, 302
  NLP and text mining techniques, 302
  personalized treatment recommendations, 307–312

Recurrent Neural Networks (RNNs), 58
Remote patient monitoring (RPM), 246–248
  challenges and considerations, 247
  in chronic conditions, 247
  efficiency and scalability, 247
  use cases, 246
Rotary positional embedding (RoPE)
  advantages of, 83
  working principle, 82

S

Social determinants of health (SDOH) analysis
  factors including, 343
  NLP and text mining techniques, 343
    via spaCy and scispaCy, 344–349
  predictive models, 343

T

Telemedicine, 8
  barriers
    cultural and organizational barriers, 249
    financial barriers, 249
    lack of interoperability and standardization, 248
    physician-centric approach, 250–251
    reliability and accuracy, 248
  benefits of, 245
  in COVID-19 pandemic, 246
  remote patient monitoring (RPM), 246–248
  service offerings, 245
Temporal association rule mining, 180
Tokenization, 36–37
Translating AI into practice
  challenges, 290

drug discovery and development, 312–321
ethical and regulatory considerations, 290
explainability and transparency
  accountability and liability, 368
  clinical decision support, 367
  counterfactual explanations and case-based reasoning, 369
  feature importance and attribution, 368
  natural language explanations and dialogue, 369
  patient autonomy and informed consent, 367–368
  patient privacy and data ownership, 374–378
  Privacy Preserving Logistic Regression model, 380–381
  PyHealth library, 378–380
  regulatory and legal compliance, 368
  rule extraction and decision trees, 368
  SHAP library, 370–374
  strategies and best practices, 369–370
medical imaging, 292–302
rare disease identification, 302–312
readmission and complication prediction
  chronic disease progression modeling, 337–343
  logistic regression model, EHR dataset, 330–337
  potential biases and limitations, 330
  real-time data, incorporation of, 329–330
  SDOH analysis, 343–349
  use of machine learning algorithms, 329

steps involved, 289
surgical planning and execution
  computer vision and machine learning algorithms, 321
  intelligent surgical robots and instruments, 322
  Medical Segmentation Decathlon, 321–329
technical barriers and interoperability issues
  data privacy and security, 350
  data quality and completeness, 349
  data standardization and normalization, 349–350
  FHIR server, 351–358
  infrastructure and computational resources, 350
  model interpretability and explainability, 350
workflow disruption and user acceptance
  bias and fairness, 361–367
  cost and resource constraints, 360–361
  liability and accountability, 359
  skill and knowledge gaps, 359
  unintended consequences and ethical concerns, 359
  user trust and adoption, 358–359
  workflow compatibility and integration, 358

U

Utility-based association rule mining, 180

V

Vanishing gradient problem, 78

www.ingramcontent.com/pod-product-compliance
Lightning Source LLC
Jackson TN
JSHW060838070925
90517JS00009B/40